Musical Sound, Instruments, and Equipment

Musical Sound, Instruments, and Equipment

Panos Photinos
Southern Oregon University

Morgan & Claypool Publishers

ISBN   978-1-6817-4680-7 (ebook)
ISBN   978-1-6817-4681-4 (print)
ISBN   978-1-6817-4682-1 (mobi)

DOI   10.1088/978-1-6817-4680-7

Version: 20171201

IOP Concise Physics
ISSN 2053-2571 (online)
ISSN 2054-7307 (print)

A Morgan & Claypool publication as part of IOP Concise Physics
Published by Morgan & Claypool Publishers, 1210 Fifth Avenue, Suite 250, San Rafael, CA, 94901, USA

IOP Publishing, Temple Circus, Temple Way, Bristol BS1 6HG, UK

*To my mother Irini, to Demetri, Marko, Ryan, Zoe,
and to Shelley who made it all possible.*

Contents

Preface

Music is a means of communication and is an integral part in many human activities. Most ancient cultures used music for celebrations, worship, healing, and preserving their stories. Archeologists have discovered that humans produced sound instruments tens of thousands of years ago. Some of our instruments today, using vibrating strings, vibrations of air columns in pipes, and vibrating rods and membranes, may have reached a mature form over two thousand years ago. Newtonian mechanics fully explained all forms of mechanical vibrations. Of course the perception of such vibrations by the human auditory system goes beyond mechanics, but it stands to reason that there are some rules connecting the mechanical characteristics of vibrations to what is perceived by our auditory system. This is the subject of psychoacoustics, which probably started when Pythagoras discovered that there are simple mathematical relations between tones that sounded pleasing, and led to the development of musical scales, and eventually what we call western music. Hermann von Helmholtz found ways to measure the frequency content of sounds, and this analysis led to better understanding of the quality of sound and of musical instruments. The advances in analog and digital electronics revolutionized many aspects of music performance, composition and enjoyment.

This book is based in part on lecture notes I prepared for my general education courses for non-science majors on the subject. I have tried to keep the number of equations and the use of mathematics to a minimum. The presentation is qualitative and does not assume technical knowledge or math skills. The content is intended primarily for a general audience and for non-science majors, and can prove very helpful for students, educators, amateur musicians, and non-specialists. I have included a brief technical section at the end of each chapter, where the interested reader can find the relevant physics and sample calculations. These quantitative sections can be skipped without affecting the comprehension of the basic material. Questions are provided to test the reader's understanding of the material. Answers are given in Appendix C. To keep in line with the objectives of the concise physics series, I kept the historical background to a minimum.

Acknowledgments

It is a pleasure to acknowledge Joel Claypool, Melanie Carlson and Brent Beckley of Morgan & Claypool Publishers for their guidance, and Jeanine Burke and Chris Benson of the IOP for their expert help in preparing this book. I thank Dr Demeter Tsounis, Mr Wataru Sugiyama, and Ms Linda Chambers for sharing their knowledge and music performing expertise, and Dr Gordon Wolfe for teaching me how to appreciate many genres. I am thankful to Mr Michali and Aleko Ieronymidi for their friendship and for sharing their expertise on instrument making. I am grateful to my wife Shelley for editing and constant encouragement.

Author biography

Panos Photinos

Panos Photinos is professor emeritus at Southern Oregon University where he has taught since 1989. He developed and taught many courses, including two courses on the physics of music. Prior to joining SOU he held faculty appointments at the Liquid Crystal Institute, Kent, Ohio; St Francis Xavier, Antigonish, Nova Scotia, Canada; and the University of Pittsburgh, Pennsylvania. He was visiting faculty at the University of Sao Paulo, Brazil, the University of Patras, Greece, and Victoria University in Wellington, New Zealand. Panos completed his undergraduate degree in physics at the National University of Athens, Greece, and received his doctorate in physics from Kent State University, Ohio. He started piano lessons at the age of 5 in Egypt, where he was exposed to a wide range of musical traditions, including Arabic, Armenian, Berber, Indian, Jewish, and Turkish. During his college years in Athens, he supplemented his income playing piano and guitar at various nightclubs in the district of Plaka, at the foothill of the Acropolis. He is a collector of traditional musical instruments. He enjoys music sessions with his family in Ashland, Oregon, and with his relatives in South Australia and his homeland, the island of Ikaria, Greece. Panos has authored over 50 research publications in scientific journals, and is the author of *Visual Astronomy: A guide to understanding the night sky*.

Chapter 1

Properties of waves

1.1 Introduction

In everyday language the term wave has several uses; for example, a wave of e-mails, a wave of enthusiasm, a wave of applause, a heat wave, and so on. The general idea is that something is suddenly going above or below normal. In more technical language, a wave indicates a repeating pattern of highs and lows in some quantity. In terms of sound, what is 'waving' is the air pressure, going higher and lower than the ambient atmospheric pressure. This chapter will introduce the basic concepts that are commonly used to characterize waves, and sound waves in particular.

1.2 Periodic waves

A most familiar wave is the pattern of circles generated by dropping a coin in a still pond, shown in figure 1.1. The pattern consists of highs and lows (crests and troughs,

Figure 1.1. A wave in a pond.

respectively) traveling outwards from the center. At each point of the surface of the pond, the water level **oscillates** in cycles, above and below the undisturbed level of the pond. We will refer to the undisturbed level as the **equilibrium** level of the water surface. The pattern propagates away from the point of impact (the source of the wave) and at each point the propagation is along the line of sight to the source. In a shallow flat-bottomed pond, the crests (and troughs) travel at the same speed. As the speed is the same for all crests, the distance between successive crests remains the same as the wave travels. The distance between successive crests (or successive troughs) is defined as the **wavelength**. The time elapsed between the crossings of two successive crests through a given point is the **period** of the wave. In one period, the wave travels a distance of one wavelength; in other words, the **speed** (i.e. distance traveled divided by time of travel) is the ratio of the wavelength divided by the period.

If we count the number of crests crossing through one point in a given time interval, say in one second, then we have a very important concept in the study of sound, namely the **frequency**. The frequency is the inverse of the period, i.e.

frequency = 1/(period)   and therefore   period = 1/(frequency).

If four successive crests cross a given point in 1 s (i.e. if the frequency is 4 crests per second) then the time elapsed between two successive crests (i.e. the period) is 1/4 of a second. As the speed of the wave equals the wavelength divided by the period, and since frequency is the inverse of the period, it follows that the speed of the wave is equal to the product of the wavelength times the frequency:

Speed of wave = (frequency) × (wavelength).

The frequency is fundamental in characterizing sound tones, and is a measure of what we call the pitch. High pitch tones correspond to high frequencies, and low pitch tones correspond to low frequencies. The frequency is measured in units of Hertz (Hz for short).

The difference in height between the top of a wave crest and the undisturbed water level is the **amplitude** of the wave. The amplitude of the wave depends on the weight and speed of the impacting object. A small coin will cause a smaller amplitude than a huge rock. The amplitude relates to the intensity of the wave. Note that as the wave pattern expands, the amplitude diminishes, and eventually the wave dies out. With sound, this observation relates to everyday experience: the farther we are from the source, the weaker it sounds.

The succession of crests and troughs in the water pond example occurs because of gravity. Water that happens to be above the equilibrium level of the surface is pulled down by gravity. The downward speed builds up, and that amount of water falls below the equilibrium level of the surface and becomes part of a trough. While moving down it pushes adjacent parts of the water upward, which become part of a crest, and so forth. The entire cycle can be viewed as an attempt of gravity to restore the water level back to the equilibrium level, as it was before the coin was dropped. In the process of restoring the equilibrium water level, it keeps overshooting the target. Thus, in our example, gravity is the **restoring force**. The overshoot occurs because of the energy imparted by the impacting coin.

All waves require a restoring force. For example, in a vibrating string, the tension of the string acts as the restoring force. There is a relation between the frequency of the wave and the strength of the restoring force. A stronger restoring force makes the up-and-down oscillation faster, which means that the frequency will be higher. This relation will be discussed in more detail in connection with strings and string instruments.

In the example of the water wave, the quantity that oscillates is the water level, as compared to the equilibrium level. In the case of sound waves in air, the oscillating quantity is the air pressure. The sound wave in air is a succession of layers of low and high pressures, the rarefactions and compressions, respectively. Low and high pressures are with reference to the undisturbed air pressure of the atmosphere. Note that, in the pond example, as the wave travels in the horizontal direction, the oscillation is up and down. In other words, the oscillation is at a right angle to the direction of travel. This is an example of a **transverse wave**: the oscillation is transverse to the direction of travel. In the case of a sound wave in air, the pressure oscillates back and forth, along the direction of travel. This is an example of a **longitudinal wave**.

It is convenient to use graphs to represent waves. The simplest periodic waveform is the sinusoidal wave, i.e. described by the *sin* (sine) or *cos* (cosine) functions known from trigonometry. Three cycles of a sinusoidal are shown in figure 1.2. The cycle or

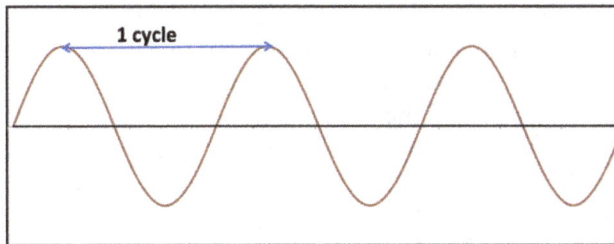

Figure 1.2. Three cycles of a sinusoidal waveform.

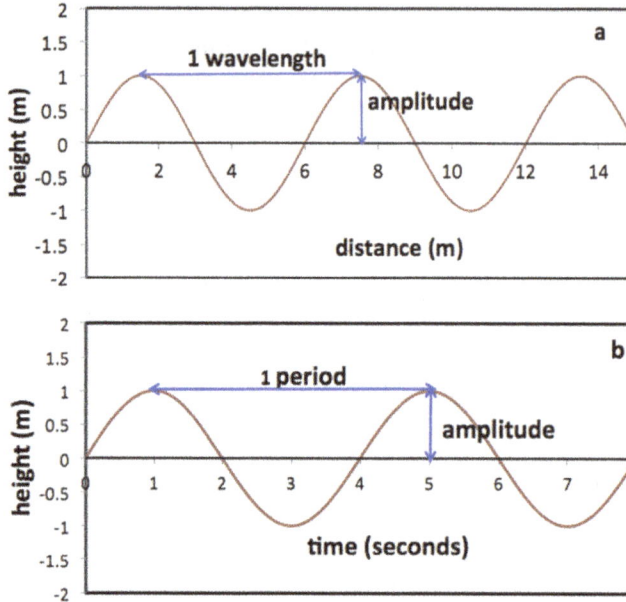

Figure 1.3. Sinusoidal waveform. (a) Horizontal axis is distance. (b) Horizontal axis is time.

periodicity of the repeating pattern is equal to the distance between two successive equivalent points; for example, two successive highs or two successive lows. The interval between adjacent highs and lows is half a cycle, and so on. The **phase** at any point of the graph is the fraction of the cycle elapsed from the starting point of the waveform. For example, in figure 1.2, the phase at the first high on the left is one-quarter of a cycle; the phase at the first low is three-quarters of a cycle.

When using graphs to represent waves, the vertical axis of the graph is used to show the value of the oscillating quantity (e.g. displacement or pressure). In the horizontal axis we usually have two options. We can choose the horizontal axis to show distance or time, as indicated in figure 1.3.

With reference to the water wave in a pond, the vertical axis shows the height of the water level. From the graph, we see that the amplitude of the wave is 1 m. If we choose to show distance in the horizontal axis, then the graph will give the profile of the height at a given instant. If the horizontal axis is chosen to indicate time, then the graph will give the variation of the height at a given point. It is important to note that the two graphs provide different information about the wave. Recalling the definitions of the wavelength and period, we see that the distance between successive crests in figure 1.3(a) is equal to the wavelength. In this example, the first crest occurs at a distance of about 1.8 m, and the second crest occurs at a distance of about 7.8 m. The wavelength is found by taking the difference of the locations of the two crests, i.e. 7.8−1.8 = 6 m. In figure 1.3(b), the first crest occurs at time = 1 s, and the second crest at time = 5 s. The interval between successive crests is the period of the wave, and in this example it is 5−1 = 4 s. As the frequency is the inverse of the period, it follows that the frequency of this waveform is (1/4)= 0.25 Hz.

1.3 Addition of waveforms

In this section we discuss simple ways in which waves can combine with each other. Comparison of the phase of the interacting waves is the key concept in understanding the outcome. Figure 1.4 shows two identical waveforms, i.e. they have the same amplitude and the same periodicity. The horizontal axis is not labeled, and can be either distance or time without affecting the conclusions. The graphs are offset vertically, and the horizontal lines represent zero displacement for each wave. To find the waveform resulting from combining waves A and B, we add the displacements at each point of the horizontal axis. The result is shown in the bottom graph, and is simply the sum of the two waves, i.e. the amplitude of the resulting wave is doubled, and the periodicity (which is the wavelength or the period depending on the choice of the horizontal axis) remains the same. In this case, we combined two waves that are in step, or **in-phase**. This means at each point of the horizontal axis, the two waves have the same phase, and the **phase difference** between them is zero.

In figure 1.5, wave B is displaced to the left by about one quarter of a cycle. In this case the two waves are not in-phase, and there is a phase difference of one-quarter of a cycle. If the horizontal axis were distance, the two crests of the two waves would be separated by one quarter of a wavelength. In the same way, if the horizontal axis indicated time, then the crests of the two waves would be separated by one quarter of the period. To find the waveform resulting from combining waves A and B, we add the displacements at each point of the horizontal axis. The result is shown in the bottom of figure 1.5. We note that the sum of waves A and B has the same periodicity, but the amplitude is smaller than the sum of the amplitudes of A and B. The crest of the resulting wave occurs somewhere in between the crests of waves A and B.

In figure 1.6, wave B is displaced to the left by half a cycle, and the result of adding these two waves is total cancellation. As the amplitudes of A and B are the same and the oscillations are in opposite directions, the cancellation is complete. If

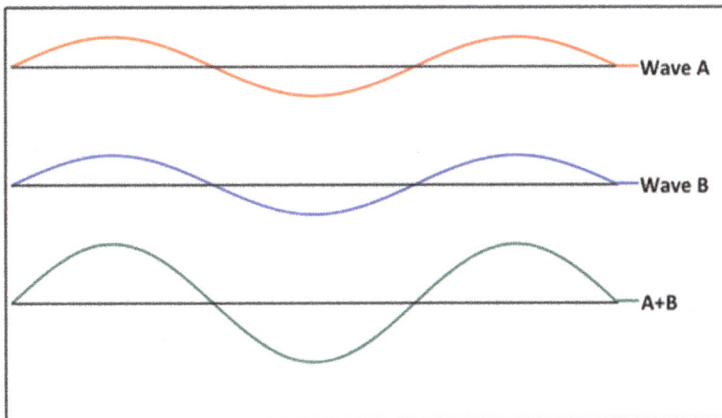

Figure 1.4. Adding two identical waveforms that are in-phase.

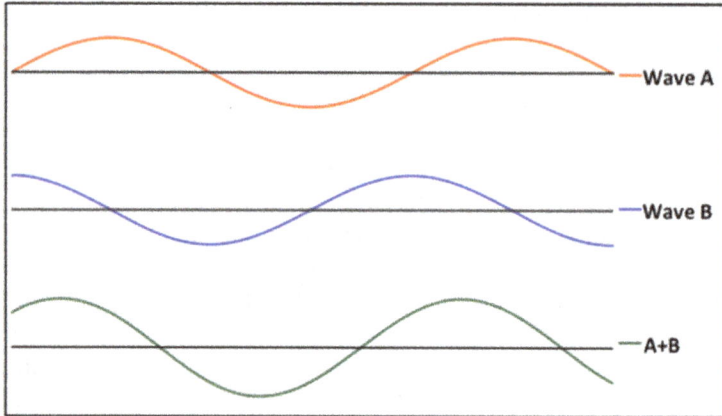

Figure 1.5. Adding two identical waveforms that are 1/4 of a cycle out of phase.

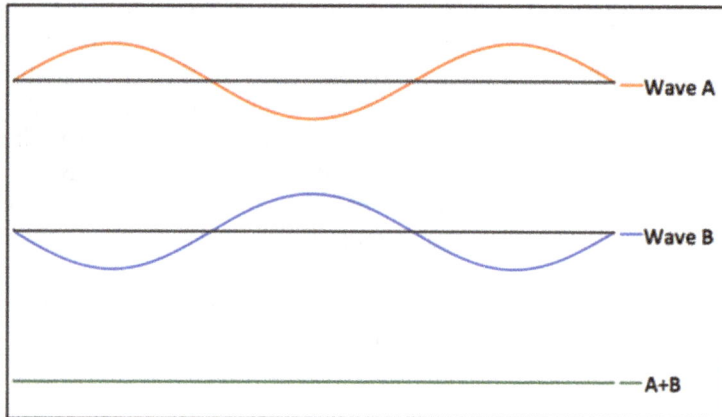

Figure 1.6. Adding two identical waveforms that are 1/2 cycle out of phase. The result is complete cancellation.

the amplitudes were not the same, the resulting wave would have amplitude equal to the difference between the amplitudes of A and B. For instance, if A had amplitude 3 and B had amplitude 1, the resulting wave would have amplitude equal to the difference $3 - 1 = 2$.

Using similar diagrams, we can add waves of the same periodicity. The general conclusions are:
- The resulting wave will always have the same wavelength and frequency.
- The amplitude will be equal to the sum of the amplitudes, if the phase difference is zero.
- The amplitude will be equal to the difference between amplitudes, if the phase difference is 1/2 cycle.

- If the phase difference between waves A and B has value between 0 and 1/2 cycle, the resulting amplitude will have value between the amplitudes A + B and A − B[1].

Cancellation of waves as shown in figure 1.6 is the basic principle behind **noise cancelling** devices. For example, noise cancellation headphones use electronics to generate replicas of the sound before it reaches each ear. The replicas generated differ in phase from the incoming wave by half a cycle. Inside each ear, the incoming sound and the out-of-phase replica cancel each other.

1.4 Beats

A very interesting situation occurs when adding two waves of slightly different frequencies. Figure 1.7 shows two waves, A and B, of equal amplitude. The horizontal axis shows time. By counting crests, we see that in the time it takes wave A to complete 10 cycles, wave B completes 8 cycles. In other words, wave A has a shorter period, therefore higher frequency than wave B. The result of adding waves A and B is shown at the bottom of figure 1.7. For our purposes, the important feature of the combined wave is that the amplitude has cycles of highs and lows, that occur approximately every 30 s. This change in the amplitude is referred to as **beats**. Thus, the amplitude of the resulting waveform A + B is **modulated**, i.e. changes with time. The period of this modulation is about 30 s. A more detailed analysis[2] shows that the result of adding two waves of different frequencies is a wave of frequency equal to the average frequency of the two waves. The amplitude of the resulting wave is not constant, but modulated at a frequency equal to the difference between the frequencies of the two waves.

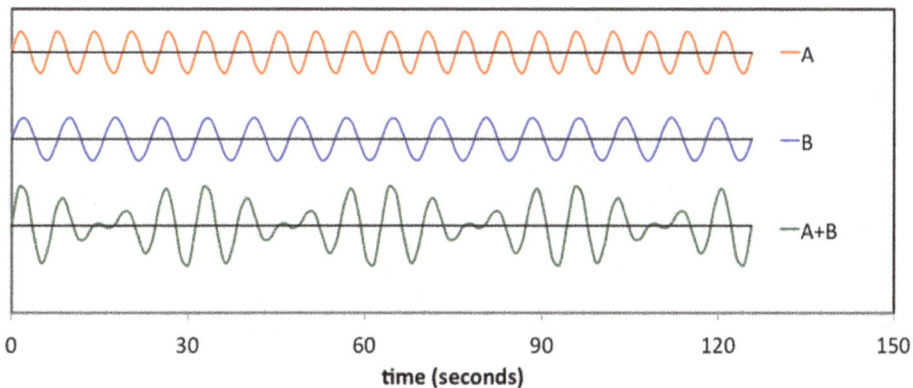

Figure 1.7. Adding two waveforms of the same amplitude but different frequencies.

---

[1] Or B − A if the amplitude of B is larger. This is so because by convention, the amplitude is always a positive number.
[2] Details are provided in appendix B.3.

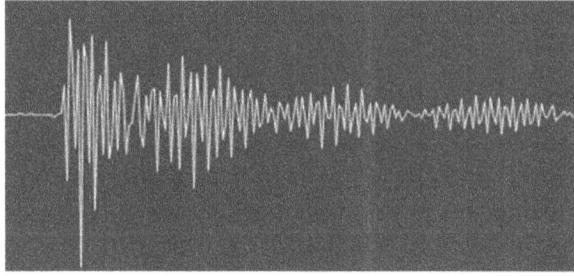

Figure 1.8. Sound produced by an out-of-tune piano. Note the four cycles of the beat pattern.

So, if we add a wave of 400 Hz and a wave of 402 Hz, the result will be a wave of 401 Hz, with amplitude that is modulated with a frequency of 402 − 400= 2 Hz, which is the **beat frequency**. Therefore, the highs of the amplitude will repeat every 1/(2 Hz) = 0.5 s, i.e. the beats will occur every 0.5 s. The phenomenon of beats plays a significant role in music and musical instruments, particularly those with double or triple strings, such as the piano. Each string in a pair should produce sound of the same frequency. If there is a mismatch of 2 Hz between the frequencies of the two strings, then from our calculation above we find that the intensity of the sound will fluctuate up and down every 0.5 s, producing a tone of poor quality.

Figure 1.8 shows the waveform of the sound produced by an out-of-tune piano. The entire waveform lasts about 2 s, from which we can estimate that the pattern of beats shown repeats about every 0.5 s. Therefore, the beat frequency is about 1/0.5 = 2 Hz, meaning that the two strings are off by 2 Hz, which, as will be discussed in section 5.6 is quite noticeable and unpleasant to the ear. Experienced tuners and string instrument players can use beats to tune their instruments very precisely.

1.5 Energy and intensity

Waves carry energy. So, when we drop a coin in a pond, some of the energy of the falling coin (the so-called kinetic energy) is transferred to the water, and carried away by the wave generated by the impact. A heavier coin or faster moving coin will result in a wave of larger amplitude, because it imparts a larger amount of energy at the point of impact. In many cases, the important quantity is not the energy imparted at the point of impact, but how much energy is *flowing* at points away from the source of the wave. To understand the concept of energy flow, it would be helpful to make an analogy to collecting rainwater in a tub. The more rain is coming down, the more raindrops will be caught in the tub, but the amount of water collected will also depend on the size of the tub because a wider tub has larger collecting area. The amount of water collected will depend on time as well: the longer we wait, the more water will be collected. We can characterize the intensity of the rain by referring to a commonly agreed upon tub area (e.g. 1 m^2) in a specific time interval (e.g. the amount collected in 24 h). Also note that the amount collected will depend on the tilt of the tub. A vertical tub will not catch much water! The maximum amount of water will be collected if the tub is perpendicular to the direction of the falling drops.

For waves, we can define the **intensity** as the rate at which energy is flowing through an area of 1 m^2. The area must be oriented perpendicular to the direction of energy flow. The rate of energy flow (i.e. amount of energy flowing per unit time) is defined as the **power**, and it is measured in Watts (W). Therefore, the intensity is measured in Watts per square meter (W m^{-2}). The intensity is related to the square of the amplitude of the wave. This means that if we double the amplitude of the wave the intensity is not doubled but quadrupled (following the square 2^2). If the amplitude is tripled, the intensity increases 3^2 = 9 times, and so on.

As the wave propagates outward from the source, the energy that started at the source is spread over a larger and larger surface. Consequently, the intensity (and amplitude) will decrease more and more the further we move away from the source. In addition, some of the energy that started at the source is absorbed, usually converted to heat as a result of friction. We refer generically to the decrease in intensity as attenuation. Attenuation will be discussed in detail in section 2.7.

1.6 Further discussion

Examples of waves

Electromagnetic (EM) waves include light, radio waves, microwaves, ultraviolet radiation, x-rays, and more. The oscillating quantity in EM waves is an electric field combined with a magnetic field. In empty space, all EM waves travel at the speed of light. EM waves can be produced by the oscillation of electrical charges.

WiFi is a wireless connection between devices based on transmitting and receiving EM waves of frequency 2.4 billion Hz (GHz for short) and 5 GHz. Similarly, an AM radio is a wireless connection, using EM waves of frequency in the range of about 500–1600 thousand Hz (kHz). And an FM radio uses EM waves of frequency in the range from 87.5 to 108 million Hz (MHz). All these waves carry the sound information, but what we hear is not the EM wave itself. The EM wave is only the carrier, the so-called **carrier wave**. The sound information is 'imprinted' on the carrier signal as a modulation of the amplitude (AM = **A**mplitude **M**odulation) or the frequency (FM = **F**requency **M**odulation) of the carrier wave. The bottom graph in figure 1.7 is an amplitude-modulated signal.

One of the most exciting discoveries in recent years was the detection of **gravitational waves**. In principle, gravitational waves can be produced by oscillations of huge amounts of mass (like stars) by analogy to EM waves that can result from oscillating electric charges. The oscillation quantity in the case of gravitational waves is the 'fabric' (or the curvature) of space, as predicted by Einstein's general theory of relativity.

1.7 Equations

Frequency, period, wavelength, and speed

The following symbols are commonly used:

c = speed
λ = wavelength
f = frequency

T = period

f is defined as $f = 1/T$

For sinusoidal waves we have:

$c = f\lambda$ and by the definition of f we have $c = \lambda/T$.

1.8 Questions

1. A person's heart rate is 120 beats per minute.
   - (a) Find the period of the heart rate in seconds.
   - (b) Find the frequency of the heart rate in Hz.
2. A wave has frequency 20 Hz and wavelength 200 meters. Find the speed of the wave.
3. Suppose we have two sound waves, A and B, traveling in air simultaneously. The amplitude of wave B is twice the amplitude of wave A.
   - (a) Which of the two waves has higher intensity?
   - (b) Which of the two waves travels faster?
4. Suppose we combine two sound waves of frequencies 100 and 106 Hz, respectively. Find the beat frequency resulting from the combination of the two waves.

Chapter 2

Propagation of sound waves

Echo by Alexandre Cabanel, The Metropolitan Museum of Art (http://www.metmuseum.org/art/collection/search/435829).

2.1 Introduction

In chapter 1, we discussed the general properties of waves propagating in a medium that is homogeneous and uniform. This is an idealized case, because our atmosphere is not uniform (it becomes thinner with altitude), and clouds make the

doi:10.1088/978-1-6817-4680-7ch2

atmosphere inhomogeneous. The general behavior of waves can be similar in many cases; for example, the path of a light wave or a sound can be bent if the air temperature along their path changes. But there are significant differences as well. For instance, one can block light by placing one's hand in front of the light source. This would not work with sound! The wavelength is the key to understanding any differences in the behavior of sound compared to light. Since behavior of light rays is easier to visualize, we will occasionally refer to the behavior of light for comparison.

2.2 Wave fronts

In order to describe the propagation of waves, it is useful to introduce the concept of **wave fronts**, which is closely associated with the concept of phase, introduced in section 1.2. We refer to the familiar example of the circular pattern created by dropping a coin in a pond, represented schematically in figure 2.1. All the crests belonging to a given circle have the same phase; in other words, they have completed the same fraction of their cycle. Compared to a given circle, the crests on the circle ahead of it have completed one cycle (i.e. one wavelength) more, and the crests on the circle behind it are one wavelength behind.

We can use a simplified picture of the wave by drawing lines connecting points of equal phase (which are circles in the case of figure 2.1) separated by one wavelength. We call these lines the **wave fronts**. We can then visualize the propagation of the wave created in a pond, as an expansion of a set of wave fronts like the red or the blue ones shown in figure 2.1. The wave always propagates in the direction perpendicular to the wave fronts, as indicated by the red arrows. In the case of the pond, we have circular wave fronts. In three dimensions, e.g. the sound from a helicopter, we would have spherical wave fronts, or a **spherical wave**. A useful idealized shape in three dimensions is the so-called **plane wave**, where the wave fronts are equidistant infinite (therefore, idealized) planes, as shown in figure 2.2.

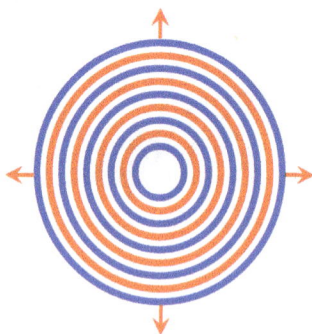

Figure 2.1. Circular wave fronts. Blue circles indicate crests and red circles indicate troughs. Adjacent crests are separated by one wavelength. The same applies to adjacent troughs. The arrows indicate the direction of propagation of the wave fronts.

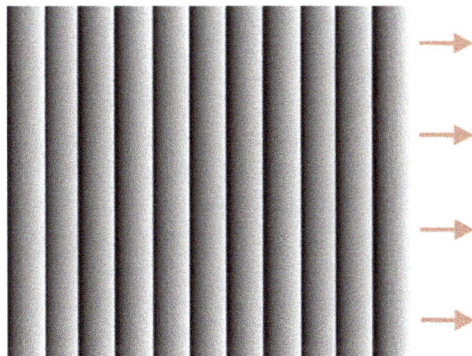

Figure 2.2. Plane waves. The arrows indicate the direction of propagation of the wave fronts.

Table 2.1. Speed of sound and density of selected materials.

Material	Speed of sound (m s^{-1})	Density (kg m^{-3})
Air	343	1.2
Water	1500	1000
Cork	400–500	200–300
Wood	3300–3600	500–800
Concrete	3200–3600	2300
Steel	6000	7800

2.3 Sound propagating in air

The most common case of sound propagation is sound propagating in the atmosphere. The speed of sound, i.e. the speed of the propagating wave fronts, depends essentially on the atmospheric temperature. The effect of other factors, such as humidity and atmospheric pressure, is minor.

At 20 °C (about 68 °F) the speed of sound is 343 m s^{-1}, or 1235 km h^{-1} (767 miles per hour (mph)). The speed of sound in air increases by about 6 m s^{-1} (about 22 km h^{-1}) for every 10 °C increase in temperature. This is equivalent to an increase of 7 mph for a temperature increase of 10 °F. For a pure tone, i.e. a sinusoidal waveform, the speed is equal to the product (frequency) × (wavelength). If the air temperature in the path of the wave increases, the speed will increase and the wavelength will increase proportionally, but the frequency will remain the same.

Sound can also propagate in liquids and in solids. Compared to the speed of sound in air, the speed of sound in liquids is higher, and the speed of sound in solids even higher. Table 2.1 lists some typical values for the speed of sound and the density of the material in kilograms per cubic meter[1]. We note that the speed of sound is higher for the denser and stiffer materials, such as concrete and steel.

---

[1] Density is the amount of mass per unit volume of material.

2.4 Reflection

The most familiar case of reflection in everyday life is the reflection of light. For example, we know that different fractions of the incident light are reflected from different surfaces. A mirror reflects almost all the incident light while a clean glass surface will reflect a small fraction (about 4%) of incident light. If the reflecting surface is smooth, the light is reflected in a specific direction, determined by the direction of the incident ray. This is the case of **specular reflection**, or simply **reflection**.

We also know that rough surfaces (for example, scratches, fingerprints, or residue on reading glasses) appear whitish. This is the result of reflections from small particles on the surface. These reflections occur in all directions, and the phenomenon is known as **diffuse reflection** or **scattering**. The smoothness of the surface is relative to the wavelength. The wavelength of light is less than 1 millionth of a meter (1 micron). Therefore, a smooth surface should have roughness (meaning, size of bumps or scratches) much smaller than 1 micron. Objects of this size are invisible to ordinary microscopes, but they can have noticeable effects. The mirrors of an amateur telescope should be smooth to at least one tenth of a micron, otherwise light will be scattered and the images become blurry.

For sound, a surface with roughness of 1 cm is sufficiently smooth[2]. This is because the range of audible wavelengths in air is about 2 cm (at the higher frequencies) to 20 m (at the lower frequencies). Therefore, surfaces with bumps or other irregularities smaller than 0.2 cm will appear rather smooth for any sound wave in the audible range, and will act as good reflectors for all audible sound waves.

In terms of the direction of the reflected wave, sound waves follow the law of reflection, which tells us that the angle of reflection is equal to the angle of incidence of the wave, as illustrated schematically in figure 2.3: the angles are measured from the normal to the wall surface at the point of incidence.

The law of reflection allows us to determine the direction of the reflected wave, but does not tell us about the amount of energy that is reflected. When a sound wave reaches the boundary of two media, for example between air and the wall in figure 2.3, some of the wave energy will be reflected back to the air, and some of the wave energy will cross the boundary, and propagate into the wall. The fraction of the energy reflected depends on the properties of the two media. The property of interest here is the so-called **acoustic impedance of the material**. The impedance is the product: acoustic impedance = (density) × (speed of sound).

For example, using values from table 2.1, we see that the acoustic impedance of air is $343 \times 1.2 = 412$ (Pa s m^{-1})[3]. Similarly, for water the acoustic impedance is $1500 \times 1000 = 1\,500\,000$ (Pa s m^{-1}), which is much larger than the acoustic impedance of air.

The percentage of the reflected energy at the boundary of two materials depends on the difference between the impedances of the two materials. If the impedance

---

[2] 1 cm = 10 000 microns
[3] Pa is a unit of pressure.

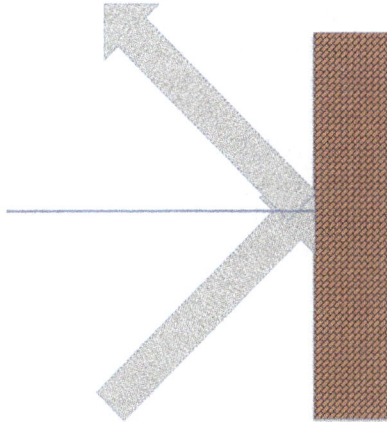

Figure 2.3. Schematic illustration of the law of reflection. The angle of reflection is equal to the angle of incidence of the wave.

difference is large, most of the energy will be reflected at the boundary. On inspection, we see from table 2.1 that the impedance of air is much smaller than the impedance of any of the materials listed in the table. Consequently, sound waves in air reaching any boundary will reflect back almost entirely, and only a small fraction will cross the boundary. For example, at the air–water boundary, about 99.9% of the wave energy is reflected back. The situation is more complicated when sound waves travelling in air reflect off a solid surface. In this case absorption occurs at the solid surface. The fraction of energy absorbed depends on the material and the texture of the surface. In general, softer materials with textured surface such as upholstery absorb significantly more than rigid smooth surfaces, such as a hardwood floor. Also, in general high frequencies are absorbed more than low frequencies. Absorption from surfaces is very important in determining the acoustic qualities of music halls and auditoriums, as will be discussed in chapter 9.

A related concept of interest to musical instruments is the impedance of a pipe. Suppose that we have a cylindrical pipe of cross-sectional area A. Then the impedance inside the pipe is

Impedance inside pipe = (density) × (speed of sound)/(cross sectional area of pipe)

If the cross-sectional area of the pipe changes, e.g. if the pipe has a constriction at some point, then there will be a change in impedance at that point, and sound waves traveling inside the pipe will be partially reflected back at the constriction. The same will happen if the pipe widens at some point. A very interesting situation occurs at the open end of the pipe. Here, the cross-section changes as well, and part of the wave that is traveling towards the open end of the pipe will be reflected back into the pipe. This happens in wind instruments such as flutes, clarinets, etc, as discussed in later chapters.

A familiar example of sound reflection is **echo**, named after a nymph in Greek mythology (shown in the chapter opening figure) who was cursed to repeat the last word she heard. We can hear our echo if we stand a distance of at least 17 m away from a wall of a large building. What we hear is the reflection of our voice. Except for special effects, echo is generally an undesirable effect in music performances. If the distance to the wall is less than 17 m, then our ear is unable to distinguish the reflection as a separate sound. In this case, the reflected sound fuses with the original sound, and the result is that the sound appears to persist for a longer time. The effect is called **reverberation**, and is noticeable in large empty rooms. Reverberation plays a significant role in the quality of acoustics in large halls, and will be further discussed in section 9.4. Echo will occur only if the reflecting surface is large compared to the wavelength of sound. We do not hear our echo from smaller reflecting surfaces; street signs, for example. The interaction of sound waves with smaller objects will be discussed below.

2.5 Interference

As discussed in section 1.3, the result of adding two sinusoidals of the same amplitude and frequency depends on the phase difference between the sinusoidals (see chapter 1, figures 1.4–1.6). The amplitude doubles if the sinusoidals are in step. If they are 1/2 of a cycle out of step, cancellation occurs. For intermediate values of the phase difference, the amplitude of the combined wave is less than the sum of the amplitudes of the individual waves. The results can be directly applied to the situation when two sound waves of the same frequency run into each other. The interaction between two waves of the same frequency is called **interference**.

Figure 2.4 is a snapshot of two waves produced by two sources that are side by side. We assume that the two waves have the same amplitude and the same frequency (same wavelength). The gray circles in figure 2.4 indicate the pressure highs (**compressions**) and the blue circles indicate the pressure lows (**rarefactions**) of the two waves. The waves have the same wavelength, therefore the distance between successive highs (which is one wavelength or a complete cycle) is the same for both waves. The same is true for the blue circles. From the above discussion, it follows that at the intersections of two gray circles the waves are in step, therefore the pressure amplitude is the sum of the two amplitudes at that point. These are points of **constructive** interference. The red dots in the figure mark some points of constructive interference. At the intersections of two blue circles the waves are also in step, therefore the amplitude is the sum of the two amplitudes at that point.

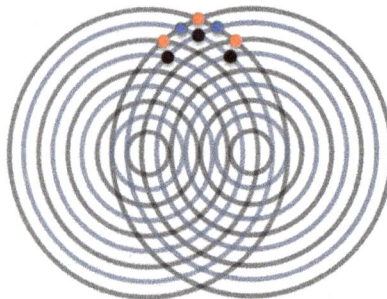

Figure 2.4. Interference between waves produced by two sources. Red and black dots mark points of constructive interference. The blue dots mark points of destructive interference.

The black dots mark some of these points. At the intersections of a gray and a blue circle, the two waves have phase difference of 1/2 cycle and the amplitude is the difference between the two amplitudes at that point, therefore some cancellation occurs. These are points of **destructive** interference, and are marked by the blue dots in the figure. At all other points in the region where the two waves overlap, the amplitude is less than the sum, but larger than the difference between the two amplitudes.

The pattern of constructive and destructive interference points shown in figure 2.4 applies to a given wavelength, which determines the spacing between the wave fronts. Similar patterns can be obtained for different wavelengths, but the points of constructive/destructive interference would be shifted accordingly. For longer wavelengths (lower frequencies) the spacing of the wave fronts will be wider and the distance between adjacent points of constructive/destructive interference will be larger. The reverse is true for shorter wavelengths (higher frequencies).

Figure 2.4 shows only a snapshot in time for a given wavelength. It is important to note that the distance of a given point from the two sources determines the phase difference between the two waves when they arrive at that point. Consequently, constructive/destructive interference will occur at the same points for that particular wavelength as time evolves. The results of interference can be noticeable in concerts, particularly with the low frequency (i.e. long wavelength) tones. A listener may hear too much of a bass tone if constructive interference for that particular wavelength occurs at the listener's location and may miss some tones altogether! Note also that destructive interference does not occur along the vertical line midway between the two sound sources. Therefore, using seats midway between the speakers would minimize the effects of interference.

2.6 Diffraction

A shadow occurs when an obstacle blocks the path of light rays. The result is more or less a two-dimensional replica of the object. With proper lighting, the shadows of ordinary objects can appear very 'crisp' indicating that light rays do not 'bend' around corners. If the dimensions of the obstacle are comparable to the wavelength of light then bending can occur. This can be demonstrated in a dark room by shining a laser pointer through a tiny pinhole on a piece of aluminum foil. If light traveled in straight lines through the hole, then one would observe a tiny spot of light the same size as the pinhole on a wall behind the foil. This is not the case actually. The spot of light past the hole is generally larger than the pinhole, indicating that light bends and spreads outward. The spreading and bending of waves around objects is called **diffraction**[4].

For example, let us consider a wave incident on a barrier that has an opening. Because of diffraction, the part of the wave that transmits though the opening will spread out at an angle. The angle is wider if the size of the opening is smaller, as illustrated in figure 2.5(b). This is the reason why we cup our hands in front of our

---

[4] F M Grimaldi apparently was the first to describe diffraction in the mid 17th century.

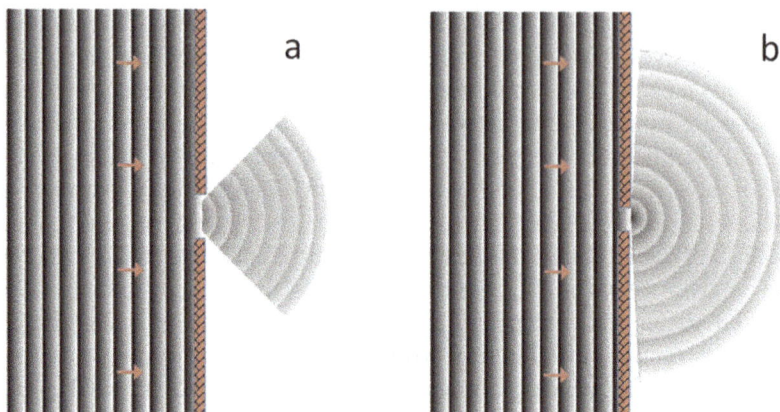

Figure 2.5. (a) Wave diffraction through an opening. (b) The spread is wider for the smaller opening.

mouth; for example, when coaches are giving instructions. Our hands create a larger opening and the sound waves (and energy) spread less. Note that if the opening becomes very small compared to the wavelength, the wave would spread completely, and if it were not for the wall, it would propagate to the left half of figure 2.5(b). In addition to the size of the opening, the amount of diffraction depends on the wavelength. If the wavelength is large (i.e. the frequency is low) the wave will spread into a wider angle. If the wavelength is short (i.e. if the frequency is high) the angle of spread will be narrower.

2.7 Absorption and attenuation

From everyday experience, we know that a 60 W light bulb may appear dazzlingly bright if it is located at arm's reach, but appears very dim if located 1 km away. Whether close or far, the 60 W rating is a property of the light bulb. It describes the rate at which energy is emitted from the light bulb, and is independent of the distance. The energy is emitted by the source at a constant rate, but as energy propagates away from the source, it spreads out in all directions. At 1 m away from the light source, the energy spreads over a sphere of radius 1 m. At a distance of 1 km from the source, the *same* amount of energy has to spread over a sphere of radius 1 km[5]. Therefore, a lot less energy crosses each square meter of the sphere at that distance, and the intensity (defined as the power crossing per unit area) is reduced accordingly.

The dependence of the intensity on the distance from the source follows the **inverse square law**, which states that the intensity depends on the inverse of the distance squared. The inverse square law applies to sound waves as well[6]. According to the inverse square law, if the distance doubles, the intensity is reduced to $1/2^2 = 1/4$.

---

[5] We neglect absorption and other losses for simplicity.
[6] A sample calculation of the intensity at various distances is given at the end of this chapter.

Table 2.2. Intensity from a small sound source of 10 W at selected distances from the source.

Distance from source (m)	Intensity (W m^{-2})
1	0.8
10	0.008
100	0.000 08
1000	0.000 000 8

If the distance is 10 times larger, the intensity is reduced to $1/10^2 = 1/100$. Table 2.2 lists the intensity for selected distances from a small 10 W sound source.

From the values in table 2.2, we see that because of the inverse square law, the intensity decreases rather rapidly. In addition to the loss of intensity due to spreading, the atmosphere may absorb part of the sound energy. The absorption affects primarily the higher frequencies over 1 kHz and increases with frequency. For example, at a distance of 10 m from the sound source, the loss to absorption is 2% for frequency 1000 Hz, 20% for frequency 5000 Hz, and 50% for frequency 10 000 Hz. As the loss due to spreading affects all frequencies more or less the same, distant sounds tend to lose the higher frequencies and sound dull. For example, a listener sitting 30 m (about 100 ft) away from the stage will lose about 50% of the 5000 Hz tones due to absorption alone.

The combined effect of all factors that reduce the intensity (including the inverse square law) is called **attenuation**. The attenuation of a wave going from point A to point B is expressed by giving the ratio of the intensities at points A and B. As the intensity can vary by factors of millions and billions, it is more convenient to use units that will keep the numbers small. These units are called **decibels**, and will be discussed in section 4.4.

2.8 Further discussion

Effects of temperature on wind instruments

As we shall see in section 6.11, the basic principle in wind instruments is that the tones produced have a wavelength determined by the length of the oscillating column of air in the instrument. Of course, as the temperature changes the length of the instrument changes, but this change is insignificant. Therefore, the wavelength of the tones produced will not change with temperature. What is significant is that the speed of sound in the instrument changes with temperature. If the speed changes, while the wavelength remains the same, the basic relation: speed = (frequency) × (wavelength) tells us that the frequency *must* change.

For example, a flute in a room where the temperature is 25 °C (77 °F) can produce a frequency of 330 Hz, which corresponds to the musical tone E4. If the temperature in the room is increased to 35 °C (95 °F) the speed of sound will increase from 346 m s^{-1} to 352 m s^{-1}, i.e. an increase of 6 m s^{-1} or 1.7%. As the wavelength remains the same, the frequency will increase proportionally, i.e. by 1.7%, which corresponds to 5 Hz. In other words, the tone will now have a frequency of 335 Hz instead of 330 Hz, which is quite perceptible. The change in temperature may not affect other instruments, for example, an accompanying guitar, which means that the flute will be playing 335 Hz, and the guitar 330 Hz corresponding to the E4. To get an idea how significant this change of frequency with temperature is, note that the next higher note (F4) has a frequency of 349 Hz, i.e. the two tones are separated by $349 - 330 = 19$ Hz. The effect of temperature on the flute is $335 - 330 = 5$ Hz, which is 26% (over a quarter of the way) of the separation between the two notes (E4 and F4), making the tones produced by a combination of the two instruments unpleasant to the ear.

Baffles

As a result of diffraction, sound waves can bend around the opening from where the sound is emerging. For example, the sound emitted from the top of a drumhead can bend around and propagate downward. Similarly, for an un-mounted loudspeaker, the sound emitted from the front of the speaker will bend around and propagate backward. In both cases we have a vibrating surface that emits sound in the forward and backward directions. As the surface is vibrating forward it creates compression in front of the surface and rarefaction in the back of the surface. A moment's reflection tells that these two sound waves from the front and the back of the surface are 1/2 cycle out of phase. If the wave from the back bends around and gets to the front side, we will have destructive interference, and the resulting sound will be weakened. This is why drums and loudspeakers, especially the ones used for low frequency (bass) sound must have a barrier, or **baffle**, to keep the sound from the backside from bending around and interfering with sound waves from the front.

Sonography

The time it takes for a wave to reflect from an object and return to the source gives us a way of measuring the distance of the reflecting object. The distance equals the speed of sound times one half of the round trip time. Methods of medical imaging use this 'echo' technique with very high frequency sound waves not audible to the human ear; so-called 'ultrasounds' to image tissues, blood vessels, organs, etc. This method is called **sonography**, and relies on the fact there is a change in the density of different parts of the body. Therefore, once the wave hits the surface separating, say, a muscle tissue from a blood vessel, the density and therefore the acoustic impedance changes, and the result is a small reflection or 'echo' of the sound wave. This will happen everywhere there is a change. By scanning a large area and converting the 'echo' times to distances, one can get a three-dimensional image of the body's interior.

2.9 Equations

The inverse square law

The mathematical equation for the inverse square law is

$$I = \frac{P_{\text{source}}}{4\pi d^2}$$

where P_{source} is the power emitted by the sound source, d is the distance from the source, and I is the intensity at a distance d away from the source.

Example: If $P_{\text{source}} = 10$ W and $d = 1$ m then $I = 0.8$ W m^{-2}

If $d = 10$ m then $I = 10/(4\pi 10^2) = 0.008$ W m^{-2}

If $d = 100$ m then $I = 10/(4\pi 100^2) = 0.00008$ W m^{-2}

2.10 Questions

1. Suppose that the wave represented in figure 2.1 has wavelength equal to 3 meters.
   (a) What is the distance between successive blue circles?
   (b) What is the distance between successive red circles?
   (c) What is the distance between adjacent blue and red circles?
2. Do you expect the echo-time to be longer in a hot desert or in Antarctica? Explain your reasoning.
3. Explain how can we overhear a conversation through an open window, even if we cannot see the people speaking. Would it matter if the people speaking were facing the window or not?
4.   (a) How long does it take for sound to travel a distance of 3.5 km?
   (b) Light travels about a million times faster. How long does it take for light to travel the same distance?
   (c) Suppose we see lightning strike and hear the thunder sound 5 seconds later. How far was the lightning?

Chapter 3

Displaying and analyzing musical sounds

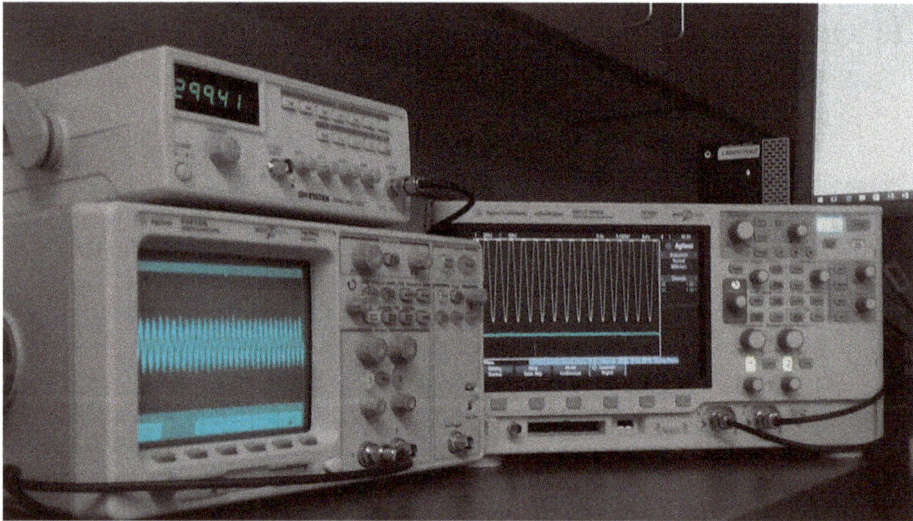

3.1 Introduction

When describing peoples' voices or sounds of instruments, we often use adjectives such as 'warm', 'crisp', 'crystalline', etc. These words may have a clear meaning, yet it is hard to quantify these characteristics, and such statements can be rather subjective. A trained ear can often distinguish the sound produced by two seemingly identical musical instruments, and in the not so distant past, instrument makers relied on sound perception alone. Today we can measure, analyze, and identify the quantities that make a voice or the sound of a musical instrument distinguishable. Analyzing a sound has many practical applications, such as voice recognition,

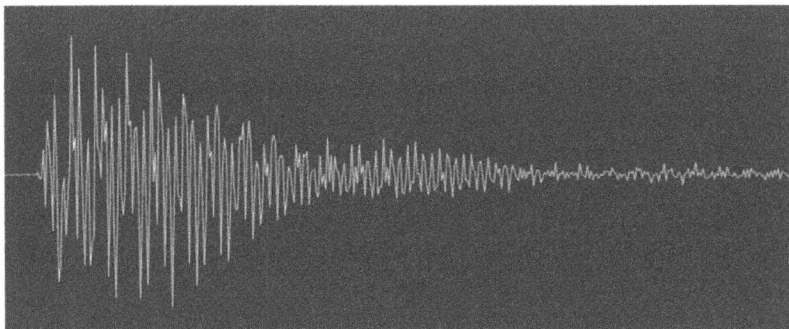

Figure 3.1. Sound signal of a single keystroke on a Yamaha keyboard.

electronic synthesizers, sound recording and reproduction, etc. This analysis requires some way of representing and displaying the musical sound.

As we will see, even the 'simplest' sounds produced by a musical instrument can appear very complicated, as illustrated by figure 3.1, which captures the sound of a single keystroke on a Yamaha keyboard. The graph shows the oscillations of an A3 note, which corresponds to a frequency of 220 Hz (see chapter 5) and lasts about 0.5 s. We note that the peaks do not have the same height. Overall, the height increases at first then dies away. We also note some 'wiggles' at seemingly irregular intervals. Are these details of any significance? The answer is *yes*, and these details contain the key information about this sound that make it different than the sound produced by striking the same note on a xylophone. In this chapter, we will consider three basic ways of representing musical sounds, and ways to extract the measurable quantities that give a tone its characteristic quality or timbre.

3.2 Measuring sound signals

Sound signals are measured using microphones (see section 8.4) and are usually expressed in units of volts (V). For example, in figure 3.1 the horizontal axis represents time. There are several choices for the vertical axis. If we are representing the signal produced by a microphone, the vertical axis will be the voltage produced from the microphone. If the signal from the microphone were amplified, then the voltage level would be different, depending on the degree of amplification. We could also convert the voltage into pressure. Remember that sound waves are essentially pressure waves. This would require of course some *calibration,* i.e. a way of converting the voltage values into pressure values. Another choice would be the intensity (see section 1.5), which would also require calibration. The meaning of all this is that the values of the vertical axis should be used just for telling what is high and what is low. Here we will use the term 'voltage', keeping in mind that the specifics of the signal level may have different units. With this understanding in mind, the result of the measurement of a sound signal is a series of voltages measured at regular time intervals.

In figure 3.1 the tone corresponds to a nominal frequency of 220 Hz, the so-called **fundamental** frequency of the tone. Obviously the graph is not a sinusoidal, in other words it is not a **pure** tone consisting of just one frequency. We know that 220 Hz means 220 cycles per second. As the graph captures about 0.5 s in time, we have about 110 positive and 110 negative peaks, i.e. 220 in total. How many data points do we need to make this graph? Well, we could start by graphing the peak values, which amounts to 220 points. By doing so we would lose the information contained in the 'wiggles' and, as mentioned above, this information is important. As discussed below, the 'wiggles' are caused by the presence of additional frequency components, besides the fundamental (220 Hz in the present example). These are the **overtones**, and have frequencies higher than the fundamental.

More voltage measurements will be required to completely capture the frequency content of this sound. For example, if an overtone has double the frequency of the fundamental, i.e. 440 Hz, we would need 220 positive and 220 negative peaks in 0.5 s, or 440 data points in total. The conclusion here is that the number of points required depends on the frequencies present in a given tone. The problem is that the frequency components of the tone can be determined only after the sound has been measured. Fortunately, we know that the range of audible sound frequencies normally does not exceed 20 000 Hz. In other words, for a tone of 0.5 s we will need at least 20 000 points. For a song that lasts 3 min (180 s) we will need $2 \times 180 \times 20\,000 = 7.2$ million points! This gives us an idea of why music files take up so much computer memory.

3.3 Visualizing a simple sound signal

To better understand the different ways of displaying a sound wave, we will start by synthesizing a complex tone using two pure tones (i.e. two sinusoidals). We will then introduce ways of displaying the parameters of the tone we synthesized. Let us consider a pure tone of frequency 20 Hz shown as wave A in figure 3.2.

The amplitude of the wave is not constant over the duration of the sound. Thus, wave A starts with amplitude 4 and at time equal to 0.15 s the amplitude decreases to 3, and again at time equal to 0.3 s the amplitude decreases to 2 where it remains

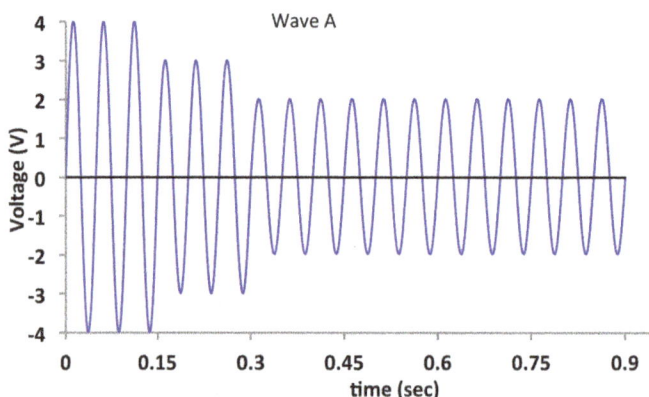

Figure 3.2. A pure tone of frequency 20 Hz. The amplitude changes at time 0.15 and 0.3 s.

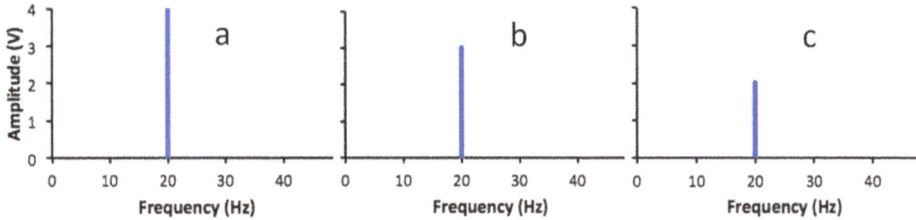

Figure 3.3. Amplitude spectrum of wave A of figure 3.2: (a) for time 0–0.15 s; (b) for time 0.15–0.30 s; and (c) for time 0.30 s and beyond.

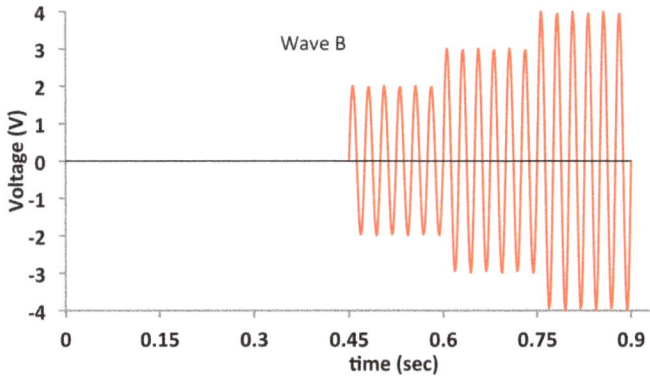

Figure 3.4. A waveform of frequency 40 Hz. The amplitude changes at times 0.45 s, 0.60 s, and 0.75 s.

until the end of the signal. The evolution of the amplitude with time is captured in the three panels of figure 3.3, which show the amplitudes for the first 0.15 s, the amplitude from 0.15 to 0.3 s, and the amplitude after 0.3 s.

With the information contained in figure 3.3, we can reproduce the tone shown in figure 3.2[1]. All one needs to do, is plot a sinusoidal wave of frequency 20 Hz, and use the amplitude values shown in figure 3.3, i.e. 4 V for the first 0.15 s, then 3 V from 0.15 to 0.3 s, and 2 V from 0.3 s on. So, essentially we have two equivalent ways of representing wave A of figure 3.2. The first way, shown in figure 3.2, is a **waveform view** (or **oscilloscope view**) of wave A. The second in figure 3.3 is an **amplitude spectrum** of wave A.

At this point we will add a second pure tone of frequency 40 Hz, shown as wave B in figure 3.4. Wave B switches-on at time 0.45 s, with amplitude 2. The amplitude increases to 3 at time 0.6 s, and increases again to 4 at time 0.75 s. Figure 3.4 is a waveform view of wave B. The waveform of the complex tone resulting from playing waves A and B simultaneously is shown in figure 3.5.

As wave B switches on at 0.45 s, the combined waveform is identical to figure 3.2 up to time 0.45 s. Consequently, the amplitude spectrum of the waveform is identical

---

[1] The significance of the phase is discussed at the end of this chapter.

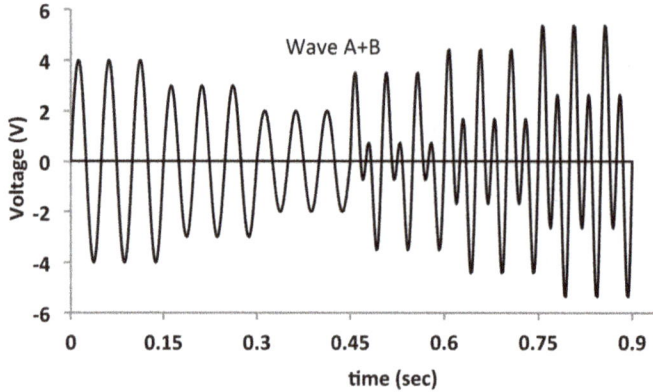

Figure 3.5. Waveform of the complex tone resulting from playing tones A and B simultaneously.

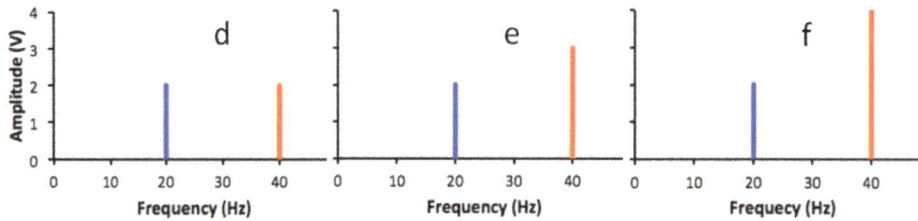

Figure 3.6. Amplitude spectrum of the complex tone of figure 3.5: (d) for time 0.45–0.60 s; (e) for time 0.6–0.75 s; and (f) for time 0.75 s and beyond.

to figure 3.3 up to 0.45 s. After 0.45 s, we note some wiggles in figure 3.5, indicating the presence of the added frequency component from wave B. Figure 3.6 shows the spectrum of the complex tone of figure 3.5 for times after 0.45 s.

We note that the spectrum for times after 0.45 s consists of two lines, one for each frequency component. The blue bars in figure 3.6 represent the amplitude of the 20 Hz tone and the red bars represent the amplitude of the 40 Hz tone. Again, using the information in figures 3.3 and 3.6 one could reconstruct the entire waveform shown in figure 3.5. Obviously, the waveform view is more compact. In our example, we have six amplitude spectra to represent one waveform view. On the other hand, we cannot readily see the frequencies involved from the waveform view.

Figure 3.7 is a so-called **sonogram** of the waveform of figure 3.5, showing the evolution of the amplitude of each frequency component. Here, the horizontal axis represents time and the vertical axis represents the frequency of each component. In the sonogram we use a color code to indicate the amplitude: orange indicates amplitude of 4 V, green indicates 3 V, and violet indicates 2 V.

The waveforms used in our example are not complicated. In real applications, the changes in amplitude may be very small, therefore the color differences may not be as obvious.

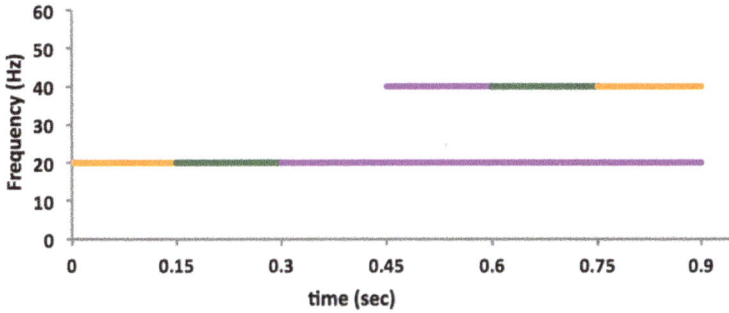

Figure 3.7. Sonogram of the waveform of figure 3.5 showing the evolution of the amplitude of each frequency component. Orange indicates amplitude of 4 V, green indicates 3 V, and violet indicates 2 V.

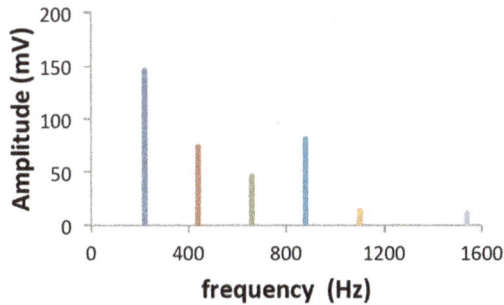

Figure 3.8. Amplitude spectrum of figure 3.1 taken 0.1 s after the beginning of the sound signal.

3.4 The spectrum of one keyboard note

In many practical situations one usually starts with acquiring the waveform view, i.e. measuring the voltage at regular time intervals. In doing so, one must make sure to get enough data points, as discussed in section 3.2. From the waveform view one could find the amplitude spectrum using Fourier analysis, which is a mathematical tool that determines the amplitude of the frequency components of a sound signal[2].

Spreadsheets and more specialized software applications offer this tool under the name FFT (fast Fourier transform) as discussed below. Applying FFT, we can find the amplitude spectrum of figure 3.1, which was produced by a Yamaha keyboard set to 'mimic' the sound of a grand piano. The results are shown in figure 3.8. The figure shows the amplitude spectrum of figure 3.1. Recall that the sound signal refers to the note A3 (220 Hz). The spectrum was taken 0.1 s from the beginning of the signal.

Note that there are several frequency components besides the fundamental frequency of 220 Hz (blue line in the figure). The additional frequency components (i.e. the overtones mentioned in section 3.2) are at frequencies 440, 660, 880, 1100, and 1540 Hz, respectively. These frequencies are integer multiples of the fundamental

---

[2] For a brief outline of Fourier transforms, see the discussion section at the end of this chapter.

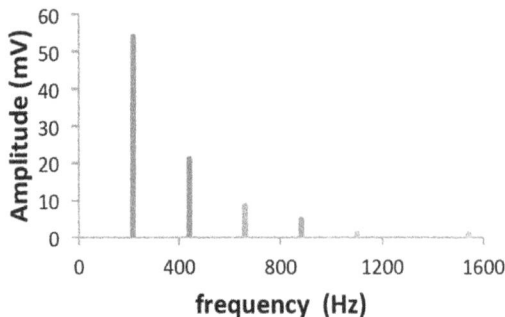

Figure 3.9. Amplitude spectrum of the waveform shown in figure 3.1, 0.2 s from the beginning of the signal.

frequency, i.e. $440 = 2 \times 220$; $660 = 3 \times 220$; $880 = 4 \times 220$; $1100 = 5 \times 220$; and $1540 = 7 \times 220$. As discussed in section 6.11, these frequencies are the harmonic overtones or simply 'harmonics' of 220 Hz.

Figure 3.9 shows the amplitude spectrum of the waveform of figure 3.1, 0.2 s from the beginning of the signal. The amplitudes of all the frequencies are smaller compared to figure 3.8, which represents the spectrum at an earlier instant (0.1 s from the beginning). The result stands to reason, as the entire signal gets weaker with time. Note also that the higher frequencies die out faster than the fundamental. This is especially noticeable for the 880 Hz (cyan line in both figures 3.8 and 3.9) which was the second highest in the earlier spectrum (figure 3.8). We conclude that for this particular sound of the Yamaha keyboard, the harmonics die away quicker than the fundamental. So as the tone is fading away, it sounds more and more like a pure tone, i.e. a tone of a single frequency (220 Hz in our example). This pattern for the harmonics is not the same for all instruments, and this is another feature, beyond the overtone content, that gives an instrument its characteristic sound quality. For example, an A3 note from a guitar, will have different harmonics, with different amplitudes, and different decay pattern for the fundamental and the harmonics, as we will see in the next section.

3.5 Comparing the sound of a steel to a nylon guitar string

As a second example, we will compare the sound of a steel guitar sting to a nylon guitar string. Here we analyze the amplitude spectrum of an acoustic guitar (see section 11.4)[3]. First, we use a steel string, and then a nylon string on the same guitar. We use the bottom string (first string) which when open normally produces a tone of 330 Hz. The strings are plucked at the same point and the spectra are taken 0.2 s after the beginning of the sound signal.

Figure 3.10 compares the sounds from the two different strings. Figure 3.10(a) shows the amplitude spectrum for the steel string. The spectrum lines show a regular spacing, at integer multiples of the fundamental frequency (i.e. 330, 660, 990, etc). We note also that the amplitude of the fundamental is not the largest of all. Instead,

---

[3] An old Chris Adjustomatic. These vintage guitars were unique in featuring an adjustable neck and adjustable bridge.

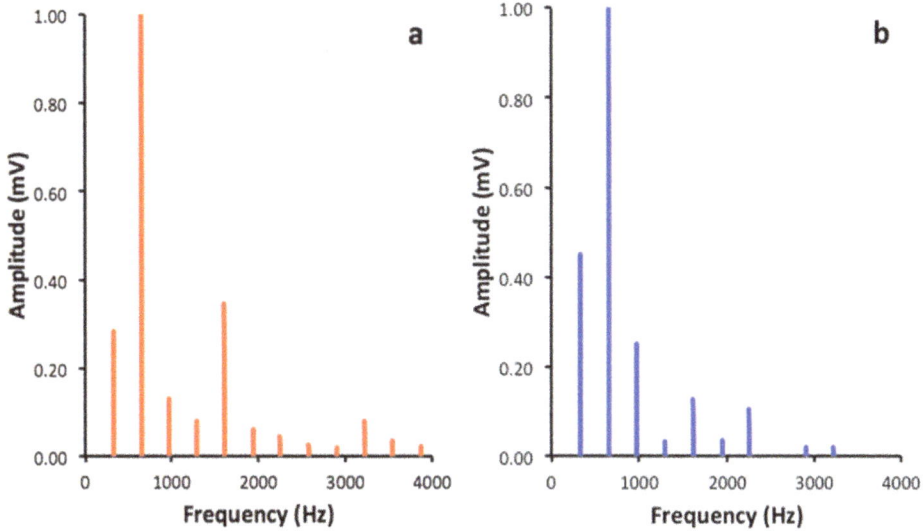

Figure 3.10. (a) Amplitude spectrum of a steel guitar string playing note E4 (330 Hz). (b) Amplitude spectrum of a nylon guitar string playing the same note.

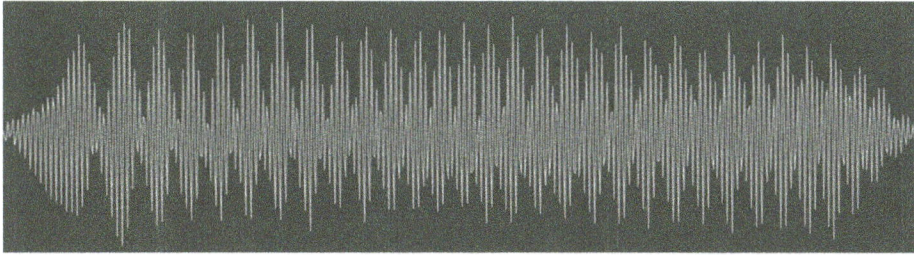

Figure 3.11. Waveform view of a C5 note (523 Hz) from a quena flute.

it is the first overtone, at frequency 660 Hz, that has the largest amplitude. This is also true of the spectrum of the nylon string, shown in figure 3.10(b). We also note that the amplitude of the fundamental is relatively larger for the nylon string. The same applies to the second overtone, at 990 Hz. This is what makes the sound quality of the nylon string 'softer'. On the other hand, looking at the higher overtones, we note that the steel string has generally more high frequency components, which gives the steel string its characteristic 'metallic' quality.

3.6 The spectrum of a flute

As a final application, we will consider a *quena*, a traditional flute from the region of the Andes in South America. Figure 3.11 shows the waveform view. The entire view is about 0.5 s, comparable to figure 3.1 for the keyboard. We note several differences; for example, the sound of the quena builds up more smoothly, sustains a steady level for a longer time, and dies out more rapidly.

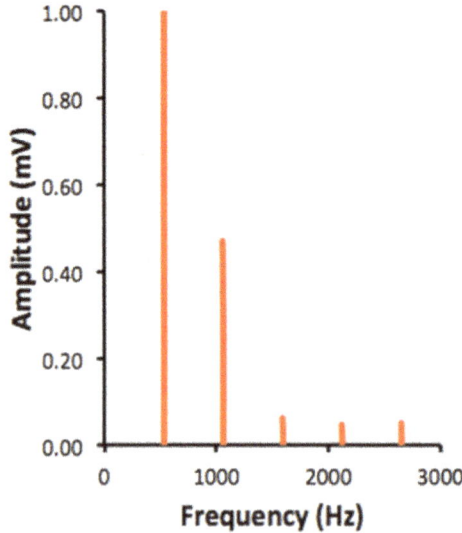

Figure 3.12. Amplitude spectrum of the waveform shown in figure 3.11 taken 0.2 s after the beginning of the sound signal.

Figure 3.12 shows the amplitude spectrum for the quena taken 0.2 s after the start of the waveform. The amplitude spectrum shows the lines corresponding to the fundamental frequency and its overtones. Here the frequency of the fundamental is 523 Hz. The lines are regularly spaced, which tells us that the frequencies of the overtones are integer multiples of the fundamental. Note also that the fundamental has the largest amplitude.

3.7 Complex tones and timbre

The sinusoidal waveform is a useful model for pure tones. As was discussed in the previous sections, the tones produced by each instrument can be expressed as a combination of pure tones. A tone consisting of many pure tones is a **complex tone**. Figure 3.13 shows how the sound of a note might begin and end. We note that the sound builds up for the first two cycles; it remains steady for the next six cycles, and then dies out. Is this a pure tone? The answer is *no*. As can be verified by looking at the symmetry of the peaks (in the figure the red dashed lines are given as a guide to the eye): the six peaks in the middle are symmetrical, as a sinusoidal *must* be. The first and second peaks are skewed to the left, while the last two peaks are skewed to the right. Therefore, they are not sinusoidal, and the same applies to the waveform as a whole. So the waveform shown in the figure is a complex tone. Comparing the waveforms of the keyboard (figure 3.1) and that of the quena flute (figure 3.11), we note a significant difference in how the waveform evolves with time. The keyboard starts with a sharp increase while the quena builds up more gradually. Tests show that if the beginning and ending parts of the waveform of an instrument were cut out, one would have trouble recognizing the instrument that produced the sound. One could verify this by playing a recorded song backwards. Therefore, we conclude

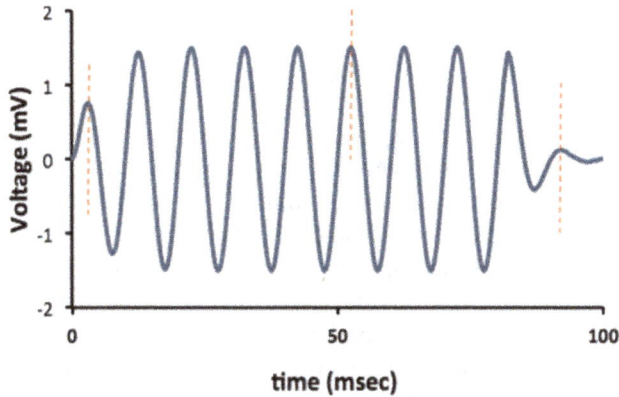

Figure 3.13. A waveform that builds up and dies out. The first and last two cycles are skewed, therefore the waveform is not sinusoidal.

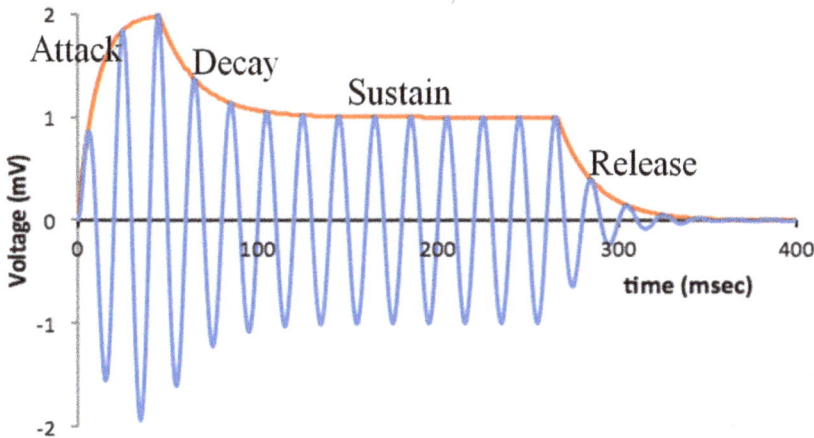

Figure 3.14. Envelope of the waveform.

that quality or **timbre** of a tone played by an instrument is determined not only by the fundamental and overtone frequencies, but also by how these frequencies build up and die out.

We could describe how the tone evolves by showing how high the amplitude gets. One way to do so is to plot the line that connects the highest points of the waveform, as illustrated in figure 3.14. The red line connects the highpoints of the oscillation, and is called the **envelope** of the waveform.

Figure 3.14 shows four features of the envelope. At the start we have the **attack**, followed by the **decay**, then comes the **sustain** part, and finally the **release**. To some extent, the performer can control these four stages of the envelope. For example, the attack stage in a guitar or violin depends on how the string is plucked or bowed. The envelope of a tone is a significant tool in synthesizing sound electronically, as will be discussed in section 11.6.

3.8 Sound analysis software applications

In the laboratory, measuring and analyzing signals, including sound signals, are accomplished using basic instrumentation, namely an oscilloscope for waveform view (or oscilloscope view), and a spectrum analyzer for the amplitude spectrum (FFT). For sound applications specifically, several apps are available online. A search for 'sound analyzer' will locate a number of apps (including open source software) suitable for use on Windows and Mac computers and laptops, and also for tablets and phones. Most of the apps record user-generated sound using the built-in computer microphone and automatically display the *sound* (which is what we referred to as *waveform view*). The signal level may be in volts, millivolts, or decibels (see section 4.4). The spectrum may also appear in the menu under FFT, and the sonogram may appear under TFFT.

Figure 3.15 shows a screen shot of AudioXplorer as an example. This particular app works for Mac computers. Here we see all three views: the waveform (top left), the amplitude spectrum (top right) taken at the location of the cursor (vertical line) on the waveform view, and the sonogram at the bottom. Moving the cursor allows a direct read-out of the voltage, time, and frequency for each view. This particular

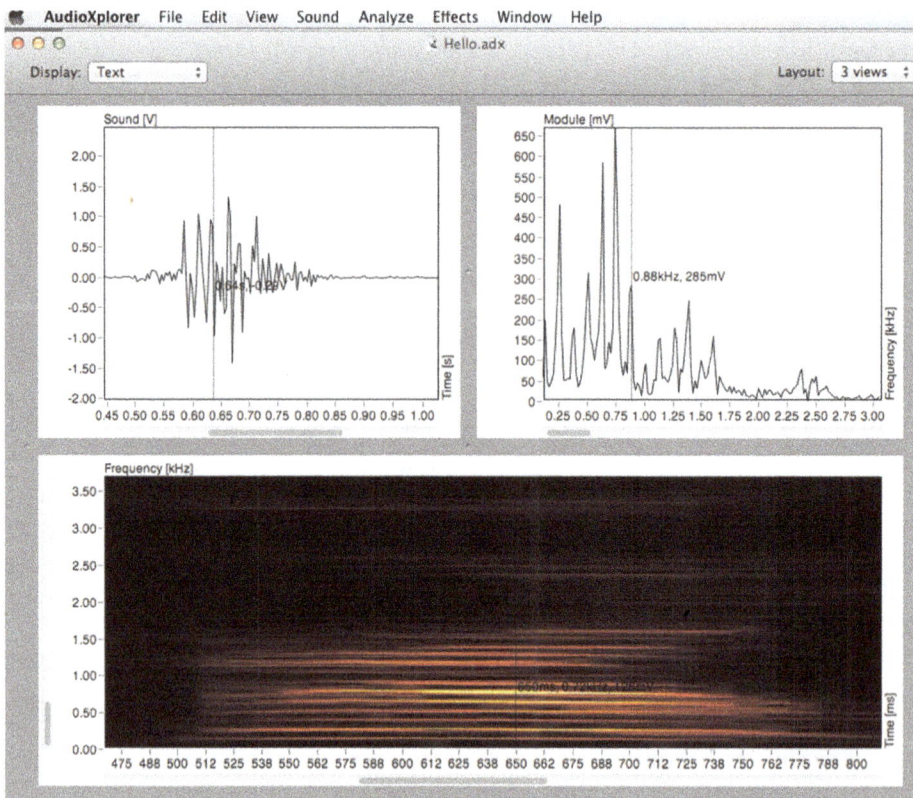

Figure 3.15. Screen shot of AudioXplorer.

app, as many others, provides a real-time option as well, which displays all three views at intervals selected by the user. In this particular example, we analyzed the word *hello*, pronounced by an adult male voice. Note that there are many frequency components extending to about 3 kHz (3000 Hz). The sonogram shows that the largest amplitudes (orange-yellow) occur in the range from 500 to 800 Hz. The voice spectrum of a child would have the larger amplitudes shifted towards higher frequencies.

3.9 Further discussion

The phase of the frequency components

The amplitude spectra tell us about the amplitude of the frequency components of the waveform. In section 1.3, we saw that when adding two waveforms of the *same frequency,* we needed to consider the phase difference between the two waves. Figure 3.16(a) shows the result of adding two waves of different amplitudes and frequencies. The frequency of the blue waveform is 50 Hz and that of the red waveform is 150 Hz. In other words, the higher frequency is an integer multiple (i.e. a harmonic overtone) of the lower frequency. The outcome of combining the two waveforms is shown at the bottom of the figure (green graph). Now suppose that we move the fundamental 1/4 of a cycle back; in other words, change the phase of the blue waveform, keeping everything else the same. The result is shown in figure 3.16(b). The resulting waveform (green graph) has a rather different appearance from the corresponding green graph of figure 3.16(a), although the pattern obviously repeats with the same period of 20 ms. What is interesting here is that the two green waveforms will usually sound the same! In other words, in general our ear is not sensitive to the phase of the harmonic components of the waveform. Therefore, having the amplitude spectrum is typically enough information to reproduce the sound of the waveform, and we do not need to worry about the phase difference between the frequency components of the waveform.

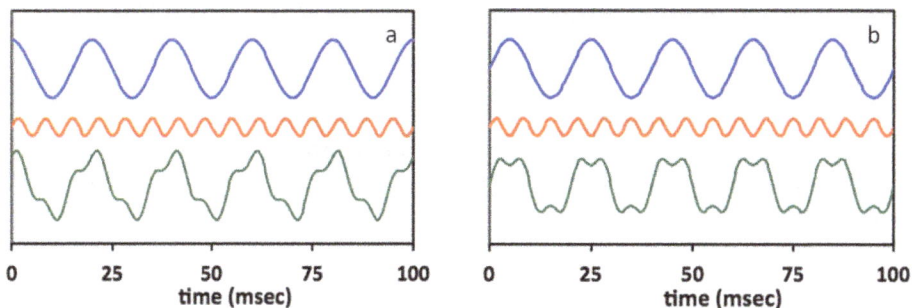

Figure 3.16. (a) Adding two waves (red and blue waveforms) of different amplitudes and frequencies. The green waveform shows the result of the addition. (b) Here, the fundamental (blue waveform) is delayed by one quarter of a cycle.

3.10 Equations

Fourier analysis

The basic idea in the analysis of sound is that any periodic function can be expressed as a sum of sinusoids. The frequencies of the sinusoids are all integer multiples of the fundamental, i.e. the frequency of the periodic function under consideration. This statement is known as the Fourier theorem. The examples of waveforms in this chapter are not periodic functions, because they have a beginning and an end, and the peak levels are not constant. If we take a 'slice' of the signal at any point in time—for example, the part of the signal contained between two successive peaks—we can create a periodic function by repeating the slice indefinitely in time. The period of this function is the time interval between the two peaks selected, and the fundamental frequency is the inverse of the period. If the time interval between the two peaks defining the slice is 0.1 s, then the fundamental frequency is $1/0.1 = 10$ Hz. Therefore, according to the Fourier theorem, the waveform for that particular time (say, the center of the slice) can be expressed by a sum of sinusoids, i.e.[4]

$$A \sin(2\pi 10t) + B \sin(2\pi 20t) + C \sin(2\pi 30t) + D \sin(2\pi 40t) + \text{etc}$$

where the remaining terms are *sine* functions with frequencies 50, 60, 70 Hz, and so on. The sum may also contain *cosine* terms following the same frequency pattern as above. The frequencies are integer multiples of the fundamental, i.e. we have harmonic overtones[5]. The Fourier theorem also prescribes the procedure to calculate the amplitudes A, B, C, etc, in the sum. We will first consider the case where the amplitudes are known.

Example: We will revisit the waveform of section 3.3, where we mixed two signals of frequencies 20 and 40 Hz, as shown in figure 3.5. If we use the interval between the last two high peaks as our slice, then our Fourier sum describing the waveform in that slice of time is

$$2 \sin(2\pi 20t) + 4 \sin(2\pi 40t).$$

If we take the slice defined by the first two high peaks in figure 3.5, the Fourier sum describing the waveform in that slice of time becomes

$$4 \sin(2\pi 20t) + 0 \sin(2\pi 40t).$$

The above two equations describe the signal in the **time domain**. The corresponding amplitude spectra (figure 3.6(f) and figure 3.3(a), respectively) describe the signal in the **frequency domain**. Therefore, the Fourier theorem is essentially a procedure to convert a signal representation from the time domain (the waveform view) to the frequency domain (the amplitude spectrum), and vice versa.

---

[4] A sinusoidal oscillation of frequency f can be described by $\sin(2\pi ft)$ or $\cos(2\pi ft)$.
[5] See discussion in section 6.11 on harmonic and non-harmonic overtones.

Calculating the amplitudes

As mentioned above, the Fourier theorem gives us the procedure to calculate the amplitudes using integral calculus. Practical applications use algorithms, which are essentially a sequence of numerical steps, to process the data points in the waveform. The FFT is an algorithm to find the amplitudes. Compared to the integral approach, the FFT calculates the amplitudes much faster and without sacrificing accuracy. As mentioned in section 3.2, our 'slice' must contain a sufficient number of data points to account for all the frequencies that may be included in the signal. As it turns out, the data must be taken at a rate equal to (or higher than) twice the value of the highest frequency component. For example, if we want to make sure that all the frequencies up to 10 000 Hz are captured, then our slice must contain 20 000 data points per second. This would be described as a '**sampling rate** of 20 kHz'. More discussion on the minimum number of data points required (or minimum sampling rate) can be found in section 7.4.

3.11 Questions

1. Suppose we want to make a waveform plot of a human voice. We know that the highest frequency contained in that voice is about 5000 Hz. What is the minimum number of points needed to plot the waveform for 1 second?
2. A sound consists of two frequencies, 100 Hz and 110 Hz. The amplitudes are 5 and 2 Volts, respectively.
   (a) Is this a pure tone? Why, or why not?
   (b) Does the waveform change with time?
   (c) What would the amplitude spectrum of this tone look like?
   (d) Does the amplitude spectrum of this tone change with time?
3. Consider the waveforms shown in figures 3.1 (keyboard) and 3.11 (flute).
   (a) Which of the two sounds is a pure tone?
   (b) From figures 3.8 and 3.9 we know that the amplitude spectrum of the keyboard changes with time. Do you expect the amplitude spectrum of the flute to change to the same degree as time elapses? More? Less? Explain your answer.
4. Examine the waveforms and amplitude spectra discussed in this chapter.
   (a) Which figures represent pure tones?
   (b) Which waveform contains the largest number of frequency components?

IOP Concise Physics

Musical Sound, Instruments, and Equipment

Panos Photinos

Chapter 4

The perception of sound

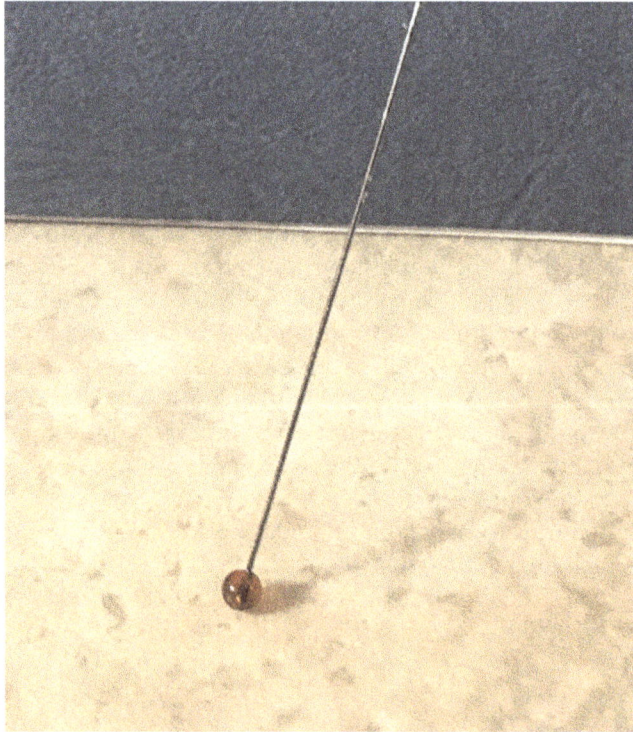

4.1 Introduction

Ordinarily sound comes to our ears from pressure waves through air. Sound waves in air are longitudinal waves consisting of alternating sequences of highs (**compressions**) and lows (**rarefactions**) relative to the atmospheric pressure. The ear picks up the pressure pattern and through complex processing turns it into what we perceive as sound. As discussed in chapter 1, a sound wave has several measurable character-istics, such as frequency, wavelength, intensity, etc. Some of the characteristics can be changed independently. For example, using a keyboard we can change the volume of a given note, hit another key to produce different frequencies, etc.

In terms of what we perceive, we can distinguish three main characteristics that correspond to the characteristics of the sound entering our ear: loudness, pitch, and timbre. There is no direct relation between the characteristics of the sound wave and the perceived sound. For example, if we double the intensity of the incoming sound wave, the loudness of the sound we perceive does not double. In this chapter we will discuss the relation between the measurable, *objective* characteristics of the sound wave and the perceived, *subjective* characteristics of sound.

4.2 Audible frequencies

The compressions and rarefactions incident on our ear stimulate the eardrum into vibration that follows the rate and pattern of the incident sound. The rate of arrival of compressions and rarefactions is what we defined as frequency in section 1.2. Therefore, the ear responds to the frequency of the incoming wave, *not* the wavelength of the sound wave. Through several stages of physiological and psychological processes, the vibrations of the eardrum are converted into what we perceive as sound.

The frequencies we normally perceive range from 20 Hz to 20 kHz[1]. This range is referred to as the **audio or audible frequency range**. In terms of frequency, sounds can be roughly classified as:

Bass	20 Hz to 250 Hz
Midrange	250 Hz to 2000 Hz
Treble	2000 Hz to 20 000 Hz

Sounds of frequency above 20 kHz are classified as **ultrasounds**, and sound frequencies below 20 Hz are **infrasounds**. Some species can hear well into the ultrasound range; for example, dogs can hear up to 40 kHz, cats up to 70 kHz, and dolphins over 100 kHz. Large animals, like elephants, can also hear infrasounds. The range of audible frequencies varies largely between individuals. With progressing age, the ability to hear high frequencies (above approximately 10 kHz) diminishes. The following discussion, unless noted otherwise, refers to young adults (18–25 years of age) with normal hearing.

An ordinary graph of the ranges of the audible frequencies is show in figure 4.1. Obviously, the low frequency range is very compressed in this graph. To get an idea of how skewed this graph is, suffice it to say that the blue range in this graphic comprises the first 39 keys (starting from left) of an ordinary piano keyboard, the green range comprises 36 keys, and the red range only 13 keys.

A better choice of the frequency axis is shown in figure 4.2, where the ranges are about equal in size. In figure 4.2, each step of the frequency axis corresponds to

---

[1] 1 kHz = 1000 Hz.

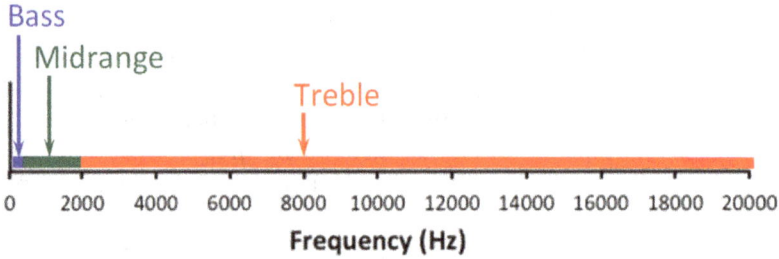

Figure 4.1. Range of audible frequencies using a linear scale for frequency.

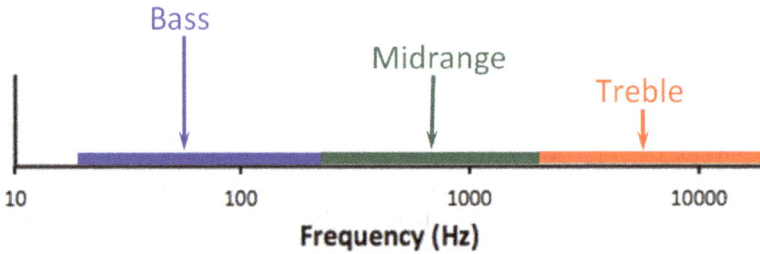

Figure 4.2. Range of audible frequencies using a logarithmic axis.

multiplying by a factor of 10. This kind of axis has a **logarithmic** scale. In figure 4.1 each step in the scale of the frequency axis represented an addition of 2000. This kind of axis has a **linear** scale. When using a logarithmic scale, in essence we are plotting the logarithm of the value, hence the name. The logarithmic scale is very convenient when graphing quantities that have a very wide range, such as the range of audible frequencies in our example.

4.3 Audible intensities

The **intensity** of a sound wave was defined in section 1.5 as the rate of energy flow through an area of one square meter[2]. The intensity is measured in units of Watts per square meter ($W\,m^{-2}$). The **threshold of hearing** (TOH) is the lowest sound intensity that can be heard by a human, on average. The TOH depends on the frequency of the sound, as discussed below. Therefore, for the TOH we need to specify both the intensity and the frequency of the sound. For a sound frequency of 1000 Hz, the TOH is about 2 trillionths of 1 $W\,m^{-2}$. In scientific notation, this is written as $10^{-12}\,W\,m^{-2}$. According to the inverse square law (section 2.7) and ignoring absorption, this is the intensity from a 10 W source set at a distance of about 600 kilometers!

The highest intensity that can be experienced without pain or damage is about 2 $W\,m^{-2}$. This is the **threshold of pain**. Higher intensities will cause severe pain and potential damage. This intensity corresponds to a 10 W source at a distance 0.6 m. We see that our hearing works with remarkably low intensities. Note that the

---

[2] We assume that the area is set perpendicular to the propagation direction of the wave.

threshold of pain is 1 trillion times larger than the TOH at 1000 Hz. In practice, it is common to use the ratio of the intensity divided by the TOH at 1000 Hz. In this form the value of the TOH is 1, and the value of the threshold of pain is 1 trillion (10^{12} in scientific notation). This is a huge range of values, and as discussed in section 4.2, a logarithmic scale would be the best choice.

When making graphs of the intensity ratio, we could use a logarithmic scale as we did with the frequency in figure 4.2, or graph the logarithm of the intensity ratio. The latter is the most common choice, because as it turns out, the human perception of measurable quantities, such as the intensity, relates more directly to the logarithm of these quantities, as will be discussed below. Therefore, it is convenient to express the logarithm of the intensity ratios in units of decibels.

4.4 The decibel (dB) unit

As the intensity of audible sounds can vary by factors of billions (10^9 in scientific notation) and trillions (10^{12} in scientific notation), it is more convenient to use units that will keep the numbers small. Such a unit is the **decibel** (dB). The dB scale essentially compares ratios. It is common to use the TOH at 1000 Hz as the denominator[3].

To express the intensity ratio of a sound in dB units, we multiply the exponent times 10.

Examples: The intensity ratio for the threshold of pain is 1 trillion (10^{12} in scientific notation). To find the dB value for the threshold of pain, we multiply the exponent of 10^{12} times 10, i.e. $12 \times 10 = 120$ dB.

Similarly, a sound intensity ratio of 1 billion (10^9 in scientific notation) is 90 dB.

To find the intensity ratio of a sound from its dB value, we divide the number of decibels by 10, and use the result as the exponent of 10.

Examples: A sound of 20 dB has intensity ratio $10^2 = 100$, i.e. the sound intensity is 100 times larger than the TOH[4].

Similarly, a sound of 60 dB has intensity ratio $10^6 = 1\ 000\ 000$, i.e. the sound intensity is 1 million times larger than the TOH.

We can use the dB values to compare sound intensities. To compare two sounds, we take the difference of their dB values, divide the difference by 10, and use the result as the exponent of 10. The sound with the larger dB has the higher intensity.

Examples: Compare the intensity of a 70 dB sound to a 30 dB sound. The difference in dB is $70 - 30 = 40$. We divide the difference by 10, i.e. $40/10 = 4$, and use the result as exponent of 10, i.e. $10^4 = 10\ 000$. Therefore, the intensity of the 70 dB sound is 10 000 times greater.

Let us compare the intensity of a 40 dB sound to a 30 dB sound. The difference in dB is $40 - 30 = 10$. We divide the difference by 10, i.e. $10/10 = 1$, and use the result as

---

[3] A more mathematical discussion on using the dB is given at the end of this chapter.
[4] The dB value of the TOH is 0, because $0/10 = 0$, and $10^0 = 1$.

Table 4.1. Decibel values and intensity ratios for selected sounds.

dB	Ratio	Example
0	1	light breathing
30	1000 (10^3)	whisper
60	1 million (10^6)	normal conversation
90	1 billion (10^9)	lawn mover
120	1 trillion (10^{12})	jet engine at 10 meters

exponent of 10, i.e. $10^1 = 10$. The intensity of the 40 dB sound is 10 times greater than 30 dB.

From the last example we see that an increase by 10 dB translates into an intensity increase by a factor of 10. For example, the intensity of a sound of 50 dB is 10 times higher than the intensity of a sound of 40 dB. Similarly, the intensity of a sound of 60 dB is $10 \times 10 = 100$ times higher than the intensity of a sound of 40 dB. In summary, every decade of decibels corresponds to a multiplication factor of 10.

Table 4.1 lists the intensity ratios and dB values for a few examples of sounds. The 0 dB represents the TOH and the 120 dB the threshold of pain.

The dB values derived from the intensity ratio represent the **sound power level**. As discussed in section 1.5, the intensity is related to the square of the amplitude of the wave. Therefore, we could use the dB scale to compare amplitudes of pressure waves rather than intensities. In such case, we speak of the **sound pressure level**. In terms of decibels the scale remains the same. For example, the difference between the TOH and the threshold of pain is also 120 dB in terms of sound pressure level. Converting dB to ratios of pressure amplitudes is slightly different than converting dB to intensity ratios. The conversion is discussed at the end of this chapter.

4.5 Threshold of hearing

Intensity is an objectively measurable quantity, and does not depend on the individual listener, the duration of the sound, or the frequency. On the other hand, what we generally refer to as 'level' or 'volume' of sound, is the perception of the stimulus, i.e. how a human perceives a sound of certain intensity, which is a subjective matter. For example, two sounds can have the same power level, yet they may *not* be perceived as being equally loud. The numbers listed here are typical values, and may change considerably depending on the individual, and the method of testing, and on the frequency used for the tests, because the sensitivity of our hearing depends on the frequency, as described below.

Table 4.2 lists representative values for the TOH for selected sound frequencies in the audible range. We note that there is a strong dependence of threshold values on the frequency. A low number of dB indicates that our hearing is sensitive to low sounds at the corresponding frequency. The highest sensitivity of our hearing is for frequencies in the range of 1000–5000 Hz. The sensitivity is lost for lower frequencies. Note, for example, that at 30 Hz we need at least 60 dB for the sound

Table 4.2. The TOH for selected sound frequencies in the audible range.

Frequency (Hz)	Sound power level (dB)
30	60
250	10
1000	0
10 000	14

to be audible. As calculated earlier, 60 dB correspond to 1 million times the intensity of the TOH. In other words, the 30 Hz sound must be at least 1 million times the TOH at 1000 Hz for the sound to be audible at all! From the values listed in table 4.2, we see that if somehow the sound power level of a stereo system is limited to 10 dB, we will not hear any the frequencies below 250 Hz, i.e. we would lose the entire bass range! Above 5000 Hz a similar loss of sensitivity occurs progressively as we approach the upper limit of the audible range. The threshold values listed above are for sound durations longer that 0.5 s. For sound durations shorter than 0.5 s, the threshold values become higher, and as discussed below, for very short durations, our hearing cannot recognize some of the characteristics of a sound.

4.6 Loudness level

Note that the above discussion refers to the TOH, which varies for different sound frequencies. For sounds above the threshold at the corresponding frequency the behavior with frequency follows the same trend. As the perceived level depends on the frequency, it is necessary to adopt a reference frequency in order to construct a loudness level scale. The unit of the **loudness level** is the **phon**. Both the loudness level and the phon are defined using 1000 Hz as reference. Thus, a sound of 10 dB (at 1000 Hz) has a loudness level of 10 phon. A sound of 20 dB (at 1000 Hz) has loudness 20 phon. In other words, the number of decibels (at 1000 Hz) equals the number of phon in loudness level at 1000 Hz. The values listed in table 4.2 represent the threshold of audible sound at the corresponding frequencies, i.e. they all correspond to 0 phon at the listed frequency.

By definition, for sounds of frequency 1000 Hz the power level and the loudness level are the same number. This is not true for other frequencies. For example, suppose that we have a 1000 Hz tone **of loudness level** 20 phon. What should be the power level of a 30 Hz tone in order to sound as loud as the 1000 Hz tone? In other words, what would be the dB value for the 30 Hz tone in order to have a loudness level of 20 phon? Measurements show that the power level of the 30 Hz sound should be about 80 dB for it to sound as loud as the 1000 Hz tone. This means that the power level difference is 80 − 20 = 60 dB, or 1 000 000 times higher! The results of such measurements are given in a set of graphs, the **equal loudness level contours**, at the end of this chapter. For our purposes, we list in table 4.3 representative values for loudness level 40 phon. We note that sounds of frequencies in the high and low end of the audible range must have very high power levels to sound as loud as the sounds with frequency 1000 Hz.

Table 4.3. Power level values (dB) required to produce loudness level 40 phon.

Frequency (Hz)	Loudness level (phon)	Power level (dB)
30	40	90
250	40	45
1000	40	40
10 000	40	55

Table 4.4. Loudness for various power levels for tones of 1000 Hz.

Power level (dB)	Loudness (sone)
40	1
50	2
60	4
70	8
80	16

4.7 Loudness

In the above discussion, we are essentially asking how much should the power level (dB) be for the sound to have the same loudness level (phon) as a given 1000 Hz sound. We can also ask another important question. Suppose that we have a tone of 1000 Hz, of a certain power level, say 40 dB. How much should we increase the power level (dB) to make the tone sound twice as loud? What sounds 'twice as loud' is again a very subjective matter, and like the loudness level and phon, quantifying **loudness** is based on averaging the results of numerous tests on many individuals. The unit of loudness is the **sone.** By definition, a tone of 1000 Hz of loudness level 40 phon has loudness of 1 sone. So, a tone of 2 sone sounds twice as loud as a tone of 1 sone. A tone of 6 sone sounds six times as loud compared a tone of 1 sone, and so on. Table 4.4 lists some values for tones of 1000 Hz.

According to the values listed, if we want to double the loudness, from 1 to 2 sone, we must increase the power by 10 dB, meaning an increase in intensity by a factor of 10. If we want to quadruple the loudness from 1 to 4 sone, then we need to increase the power from 40 to 60 dB, an increase of 20 dB, which means increasing the power 100 times. The practical aspect of these comparisons is that two voices singing together do not sound twice as loud. It may take ten voices to sound twice as loud as one voice. The values listed here apply to 1000 Hz tones. Similar, but quantitatively different, results are obtained for other frequencies as well[5].

In summary:

- The power level (in dB) tells us about the intensity of the sound, independent of the frequency.

---

[5] See discussion at the end of chapter on how these numbers are calculated.

Table 4.5. Values of the just noticeable difference for power levels 30, 60, and 80 dB.

| Frequency (Hz) | Just noticeable difference in intensity (dB) | | |
	30 dB	60 dB	80 dB
250	1.2 (30%)	0.5 (11%)	0.4 (10%)
1000	1 (25%)	0.4 (10%)	0.3 (7%)

- The loudness level (in phon) compares two tones of different frequencies. It tells us how many dB should each have so that the two tones are perceived as equally loud.
- The loudness (in sone) tells us how much we should increase the loudness level (phon) of a tone if we want to make it sound 2, 3, 4, etc times louder.

4.8 Just noticeable difference

Related to the discussion of the previous section is the smallest *change* in power level that can be perceived. This is usually referred to as the **just noticeable difference** in sound power level. As noted in the preceding discussion, the difference depends on both frequency and power level of the tone. Listed in table 4.5 are representative values of the just noticeable difference.

From the values listed, we note that at low levels (30 dB) the change must be about 1–1.2 dB or 25%–30% (values in second column) to produce a noticeable change in loudness, compared to 0.3–0.4 dB (7%–10%) at the higher levels. The values suggest that our ability to distinguish changes near the TOH is not very good. The same actually applies at very high levels (around 100 dB and above). Our ability to distinguish at intermediate sound levels improves, as indicated by the lower values of 0.3–0.5 dB (7%–11%) listed in the last two columns.

4.9 Masking

Our discussion so far has been focused on sinusoidal sounds consisting of a single frequency. These sounds are **pure tones** and are of great importance in the laboratory environment. For example, tests of the TOHs are done with pure tones. When two or more pure tones are present simultaneously, the situation is more complicated, and in general the response depends on the frequencies of the tones present. For example, if the frequencies of the two tones are close, then some **masking** occurs. If we check the hearing threshold at 250 Hz (the testing frequency) with the simultaneous presence of a 240 Hz tone (the masking frequency) then the TOH for the 250 Hz tone will be higher than the 10 dB value listed in table 4.2. On the other hand, if we check the threshold for 250 Hz (the testing frequency) with the simultaneous presence of a 10 000 Hz tone masking frequency, the threshold we find for 250 Hz tones will be 10 dB, i.e. the same as it is without the presence of the 10 000 Hz masking frequency. How close the frequencies have to be for masking to occur depends on the testing frequency, and the difference between the testing frequency and the masking frequency. In addition, the extent of masking depends on the intensity of the masking frequency. Higher masking intensity makes the threshold for the test sound higher.

4.10 Frequency and pitch

Loosely speaking, the pitch of a sound wave refers to the frequency of that sound. The frequency refers to the stimulus. The frequency can be objectively measured in Hertz, and does not depend on the subject. The pitch is the perception of frequency by the individual, and as such, depends on the individual being tested, and how the test is conducted. Some individuals are able to recognize or sing a musical tone or note of a given frequency. Also, starting from a given note, many individuals can recognize or sing the next (higher or lower) note, and thus can carry a tune very well. But most people would have difficult identifying a pitch that is exactly midway between a high (say 3000 Hz) and a low pitch (say 60 Hz).

Recall from section 4.8 that in distinguishing sound power level changes, our hearing cannot perform to better than a few percent. In contrast, our hearing is much more sensitive to frequency differences. Table 4.6 lists representative values of the **just noticeable difference** in frequency, for selected frequencies in the audible range. In terms of frequency (middle column) the **just noticeable difference** increases with frequency. Note, however, that compared to the values in table 4.5, the % values in the third column indicate that our hearing is much better at distinguishing frequency changes than distinguishing power levels.

It is interesting to note that when more than one frequency is present, our hearing may perceive additional frequencies, called combination frequencies, that are not included in the sound signal. For example, if we sound two tones, say 220 Hz and 320 Hz, we may hear a tone of 100 Hz. The frequency of the additional tone is the difference $320 - 220 = 100$ Hz, and hence it is called the **difference tone**. Another interesting situation occurs when we have a sequence of frequencies say 330, 440, 550, and 660 Hz. Note that the frequencies are multiples of 110 Hz; in other words the frequencies are overtones of 110 Hz. In this case, the listener will perceive the combination as a pitch of 110 Hz, the so-called **virtual pitch**.

In section 3.5, we compared the sound frequency spectrum of guitar strings (see figure 3.10). What is noteworthy is that the lowest frequency, which we associate

Table 4.6. Just noticeable difference at various frequencies.

Frequency (Hz)	Just noticeable difference in frequency	
	(Hz)	(%)
100	1	1
500	2	0.4
1000	3	0.3
4000	20	0.5

with the pitch of the tone, is not the loudest. Yet, because of the virtual pitch, our hearing associates the pitch with the lowest frequency. Another interesting application of the virtual pitch is the perceived bass sound from small loudspeakers and earphones (see section 8.7). Our ears may pick up a sequence like 220, 330, 440 Hz, and perceive a 110 Hz pitch which actually may be absent from the output of the loudspeaker altogether.

4.11 Critical bands

In terms of musical sound, an important question is whether we perceive two *simultaneous* pure tones as two separate tones or one tone. This answer depends on the frequency and the frequency difference between the two tones. For large differences in frequency, the two tones are perceived as separate. If the two frequencies differ by a few Hertz, the phenomenon of beats occurs, as described in section 1.4. Beats are clearly noticeable if the beat frequency (i.e. the difference between the frequencies of the two tones) is less than about 10 Hz. For slightly larger frequencies, the two tones are not resolved, and the combination produces an unpleasant **roughness** in the sound. Roughness occurs when the two frequencies are within the same **critical band** of frequency. For example, for tones around 100 Hz, the critical band extends to 145 Hz. So, if we play *simultaneously* a tone that is within the critical band of the 100 Hz tone, say a tone of 120 Hz, roughness will occur. For frequencies around 1000 Hz, the critical band extends to about 1150 Hz, meaning that the combination of 1000 Hz and 1030 Hz is a rough sound. The combination of tones outside the corresponding critical bands may or may not sound pleasant, as discussed in section 5.6.

4.12 Hearing, vision, and the role of the brain

In this section, we briefly discuss some of the common features of our hearing and vision. In both cases, the stimulus is a wave. Sound waves are longitudinal, but light waves are transverse (see chapter 1). Both hearing and vision respond to the frequency of the wave, but the frequencies of sound waves are billions of times lower compared to the frequencies of light waves. The range of audible frequencies is from 20 Hz to 20 000 Hz. Comparing the low and the high end, we see that they differ by a factor of 1000. The range of visible frequencies is from 38 to 77 trillion Hz. Comparing the low and the high end, we see that they differ by a factor of two. So, the range of audible frequencies is relatively much wider.

In terms of intensity, our hearing range is about 120 dB, and as explained above corresponds to a factor of 1 trillion. Our vision responds to a slightly wider range of intensities, about 10–100 trillion. In both cases, the response to intensity is not direct. Doubling the intensity of sound does not double the perceived loudness. Similarly, doubling the amount of light shining on a surface does not double the perceived brightness. Our hearing can pick up small differences in frequencies, but is not as good at picking up differences in intensities. The reverse is true for vision: we can pick up slight differences in light intensity, but we cannot pickup slight differences in frequency (color). Our vision has a unique ability to combine light of different colors (i.e. different frequencies) and interpret the combination as a different color. For example, all the colors seen on a computer display are generated using red, green, and blue pixels only. This is the basis of the so-called RGB display. Thus, combining green and red pixels can give us a series of yellow-orange hues. Although the underlying mechanism is very different, combining sound frequencies can also produce analogous effects, such as the difference tone and virtual pitch described in section 4.10.

An important common feature is that both systems have two 'receivers'. We have **binocular** vision (two eyes) and **binaural** hearing (two ears) which enhance the perception of our surroundings. Our field of view is about 200 degrees wide. Our hearing has a continuous field of 360 degrees. Binocular vision allows us to determine depth. Binaural hearing allows us to determine the direction and motion of the sound source, even when the source is not visible. This is referred to as **localization** of the sound source.

For low frequency sounds, under 1000 Hz, localization relies on the difference between times of arrival to each ear. If the sound source is to our right, the sound arrives to our right ear first, and then to the left ear. For higher frequencies, localization relies on the difference in intensity. If the source is to our left, then the sound intensity at our left ear is higher. The localization cues given by binaural hearing are often used in filmmaking. Using a stereo sound system, the sound engineer can progressively shift the volume of a sound source, e.g. a person speaking, from the right to the left channel. The localization cue to the viewer is that the speaker is moving, even if the person speaking is not visible.

Our hearing handles less information than our vision. The ear receives a time sequence of pressures, while our vision receives two-dimensional images. Therefore, the number of sound-detecting cells in the ear is relatively small (under 10 000) while our photoreceptors inside the eye exceed 100 million. In either case, the bits of information received would overwhelm the brain. This does not happen, because the information is locally processed first (i.e. inside the ear or eye) and the brain receives a 'summary' of what is important.

4.13 Further discussion

The equal loudness contours

ISO is the International Organization for Standardization. Graphs representing the loudness-intensity relation based on many independent tests are shown in figure 4.3, and represent the 2003 revision of older ISO standards.

The subjects were 18–25 year olds. The sound source produced pure tones and was directly in front of the listener, so the test was using both ears (binaural listening). In the graph, we use the power level on the vertical axis. Note that the vertical grid lines are not equidistant, and the value of the steps in the horizontal axis (frequency) change by factors of 10. The vertical grid lines to the right of 10 mark the values 20, 30, 40 Hz, etc. The lines to the right of 100 mark the values 200, 300, 400 Hz, etc. The red curves indicate equal loudness level, and are shown in steps of 20 phon. The red curve marked 40 shows the dB intensity value that would be perceived as having loudness level equal to 40 phon.

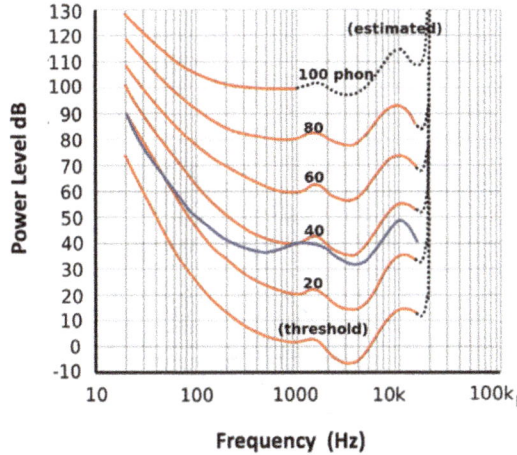

Figure 4.3. Equal loudness contours (image credit: Lindosland).

Suppose we have a 1000 Hz tone of power level 40 dB. What should be the power level of a 100 Hz tone in order to sound as loud as the 1000 Hz tone? Using figure 4.3, we look at the 40 phon curve at 100 Hz and read the power level from the vertical axis, which is 62 dB in this case. Similarly, we find that for a 10 000 Hz tone to sound as loud as the 40 dB 1000 Hz tone, we see that the 10 000 Hz tone should have power level of 56 dB. The frequencies at which our hearing requires the least power level (i.e. our hearing is the most sensitive) correspond to the lowest values of the red curves, and occur around 3000 Hz (3 kHz). The blue line indicates the older standard values at a loudness level of 40 phon.

Some cues from hearing

Besides the cues discussed in section 4.12, the frequency of a sound can give us some distance cues. As discussed in section 2.7, for complex sounds traveling through the atmosphere, the high frequency components are attenuated more than low frequency components. Therefore, a complex sound from a distant source will lack the high frequencies, and sound 'dull' compared to a nearby source. Also, the frequency of a tone, as received by an observer, can be different depending on the motion of the sound source. A most familiar example is the frequency of the sound we receive from the siren of a fire truck. When the siren is approaching, the frequency is higher. When the siren is moving away, the frequency is lower. Of course, for the driver the frequency remains the same. This is an example of the so-called **Doppler effect**. Note that the motion affects the frequency reaching us. Therefore, this is an objectively measured shift in the frequency, and is not associated with perception. In the case of light waves, the shift in frequency is extremely small, therefore at ordinary speeds our vision cannot detect the Doppler effect due to the motion of a light source.

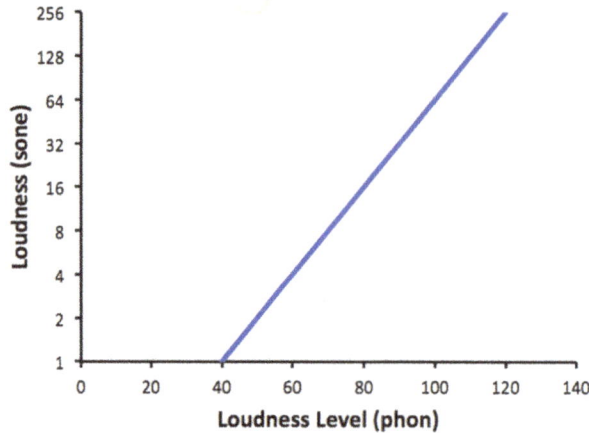

Figure 4.4. Loudness (in sone) versus the loudness level (phon).

Relating sone, phon and dB

In section 4.7, table 4.4, we listed values of loudness and power level for 1000 Hz. Figure 4.4 shows a graph of the loudness (in sone) versus the loudness level (phon).

Note that the vertical axis is logarithmic and the steps are multiplicative factors of two. For loudness levels below 40 phon, the relation is not linear, and not shown here. We could translate the loudness level (phon) into power level (dB) using figure 4.3. In doing so, one has to keep in mind the frequency dependence. For the special case of 1000 Hz, the situation is very simple, because the numerical value of the loudness level and power level are equal by definition. This was the case used in table 4.4 for simplicity.

The duration of sound waves

The process of recognizing the characteristics of a sound wave begins in the inner part of the ear, and it takes some time to sort out the information content of the sound wave and eventually transmit the signals by the acoustic nerve to the brain. The models that explain the perception of sound are complicated and beyond the scope of this book. Nevertheless, it is plausible that if the duration of the sound wave is short compared to the time needed to complete the perception process, the brain will be unable to completely identify the characteristics of the incoming sound wave. Also, if the duration of the wave is too short, the energy accumulated in the inner part of the ear may not be sufficient to properly stimulate the acoustic nerve. The result is that our hearing cannot recognize sound waves of short duration, especially if the intensity is low. Obviously, we need at least one cycle (1/frequency) to recognize the frequency of the wave, so the duration of a recognizable frequency cannot possibly be shorter than the time it takes to complete one cycle. For example, if the wave frequency is 1000 Hz, then it takes $1/1000 = 0.001$ s for a complete oscillation cycle. For a frequency of 250 Hz, it takes $1/250 = 0.004$ s to complete one cycle. The number of cycles required depends on the frequency as well. It takes about ten cycles for a frequency of 1000 Hz to be clearly recognized, and in terms of duration $10 \times 0.001 = 0.01$ s. This is the shortest duration that a sound of 1000 Hz can have, and be recognizable. For a frequency of 250 Hz it takes about four cycles. Therefore, the shortest duration a 250 Hz sound can have and be clearly recognizable is $4 \times 0.004 = 0.016$ s.

4.14 Equations

Converting dB to ratios

The dB unit allows us to compare two intensities, using the logarithm (base 10) of their ratio.

Example: If $I_1 = 120$ (W m^{-2}) and $I_2 = 40$ (W m^{-2}) the dB value is

$10 \times \log(I_1/I_2) = 10 \times \log(120/40) = 10 \times \log(3) = 10 \times 0.477 = 4.77$ dB.

In other words, I_1 is 4.77 dB higher than I_2. Here we used I_2 as reference. We could have used I_1 as reference:

$10 \times \log(I_2/I_1) = 10 \times \log(40/120) = 10 \times \log(1/3) = 10 \times (-0.477) = -4.77$ dB.

In other words, I_2 is 4.77 dB *lower* (because of the '−' sign) than I_1. If we know the dB value, we can find the ratio, by inverting the logarithm.

Example: If I_1 is higher than I_2 by 3.4 dB, then by the definition of the dB
$10 \times \log(I_1/I_2) = 3.4$ dB or

$\log(I_1/I_2) = 3.4/10 = 0.34$ therefore,

$(I_1/I_2) = 10^{0.34} = 2.19$.

We now know the ratio. If we know the value of the reference (for example, if we know I_2) then $I_1 = 2.19 \times I_2$.
Or if we know I_1, then
$I_2 = I_1/2.19$.

The essential point is that the dB is a relative scale. We cannot get from the dB the values of *both* intensities, we can only get their *ratio*. Of course, if we know the ratio, and one of the two intensities (the 'reference' which is usually the denominator in the ratio) then we can find the other (the numerator). Negative dB values indicate that the denominator is larger than the numerator, i.e. that ratio is less than one.

The intensity is related to the square of the amplitude. When comparing **amplitudes**, such as the amplitude of the pressure of a sound wave, the decibel value is

Amplitude level (in dB) $= 20 \times \log(A_1/A_2)$.

4.15 Questions

1. From table 4.1 we see that the sound power level of a whisper, a lawnmower and a jet engine at 10 meters are 30, 90 and 120 dB, respectively.

(a) Does this mean that a whisper plus a lawn mower (30 + 90 = 120) are as loud as a jet engine?

(b) From the intensity ratios, we see that the intensity of a whisper is 1000 times higher than light breathing, meaning that it takes 1000 people breathing lightly to sound as loud as one person whispering. Using the same logic, find how many lawnmowers it takes to sound as loud as a jet engine.

(c) What intensity ratio corresponds to power level of 40 dB?

(d) Suppose we have 10 lawnmowers running. What is the power level of their combined sound in dB?

2. Suppose that we have two sound sources producing pure tones of frequency 30 and 1000 Hz, respectively.

(a) If the two sources are adjusted to output a power level of 5 dB each, which source will emit more power?

(b) If the two sources are adjusted to output a power level of 5 dB each, which source will sound louder?

(c) Does sound power level and loudness level mean the same thing? If not, then explain the difference.

(d) If the two sources are adjusted so that the perceived loudness level from each source is 40 phons, which source will sound louder?

(e) How much should the power level of the 1000 Hz source be increased in order to make it sound twice as loud?

3. For pure tones of about 1000 Hz, the just noticeable difference in frequency is about 3 Hz.

(a) Suppose that one hears a tone of 1000 Hz, followed by a tone of 1002 Hz. Would that person be able to tell the difference in pitch?

(b) Suppose that the two tones are sounded together. Would one be able to tell that the two tones do not have the same pitch?

(c) What is different between the situations described in parts (a) and (b) of this question?

Chapter 5

Musical scales and temperament

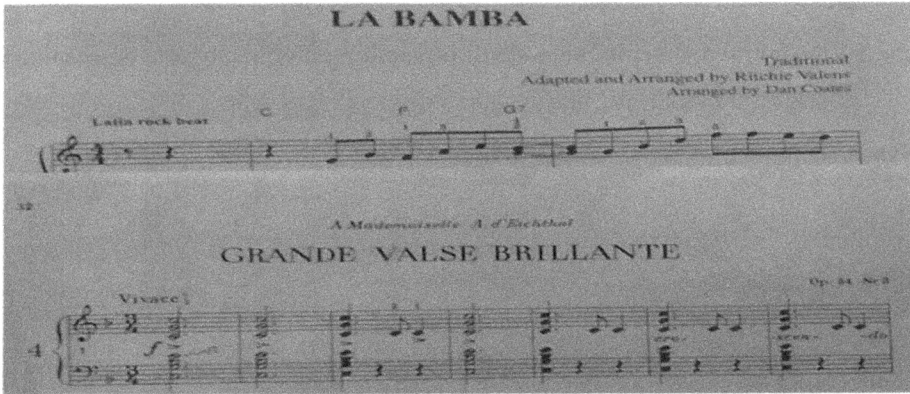

5.1 Introduction

The range of audible frequencies is essentially a continuum, a lot like the colors of the rainbow. So in principle one could use any combination of frequencies simultaneously or successively, in a similar way that an artist uses colors. However, the results of these approaches may not be pleasing to the ear or the eye. Music is based on a set of tones of interrelated frequencies and on rules governing how these tones should be combined together to produce a pleasing result. Today we can use scientific arguments to demonstrate the plausibility of these interrelations and rules. It is clear however that the development was driven primarily by aesthetic criteria with influence from various cultures, as is obvious in jazz, flamenco, etc. The term **scale** derives from the Latin *scala*, meaning staircase. Therefore a scale is a set of *frequency* steps. In this chapter we describe the frequency relations between the steps of musical scales.

5.2 Keyboard notes

To make the description more specific we will refer to the keyboard. A standard piano has 88 keys. Figure 5.1 shows a section of a piano keyboard. The keys are labeled with their English names[1].

At once we note that the pattern is repeating, and that the black keys are labeled with reference to the white keys. For example, the black key between C and D is labeled as C♯ (pronounced C sharp) or D♭ (D flat). We also note that there are 11 different keys (counting black and white) between two successive As. If we think of the keys as steps in a ladder, then there are 12 steps between successive As. The same is true for successive Bs, Cs, etc. It is common to refer to the C near the middle of the piano keyboard (fourth C from left end) as **middle C** or C4. All the keys can be numbered similarly.

Figure 5.2 shows a section of the keyboard starting from C2. Note that the numbering is based on the Cs not the As. Therefore, the white keys between C4 and C5 are D4, E4, F4, G4, A4, and B4.

The black keys are numbered following the same scheme. To describe the direction of movement on the keyboard we will use the term **ascending** to indicate the direction from left to right (which is the direction of increasing tone pitch) and **descending** to

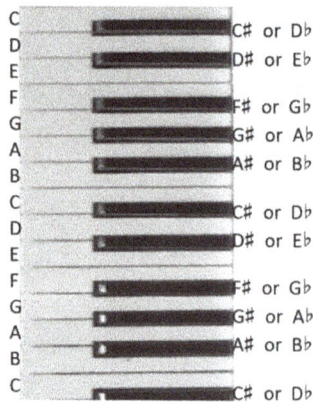

Figure 5.1. A section of a piano keyboard.

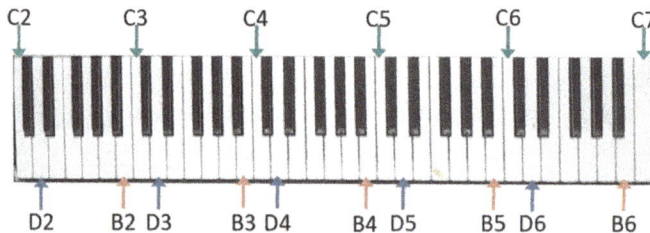

Figure 5.2. A section of the keyboard starting from C2.

---

[1] Many countries use the **solfège** system, corresponding to C(do), D(re), E(mi), F(fa), G(sol), A(la), B(si).

indicate the direction from right to left. Note that some of the white keys are separated by a black key (e.g. C4 and D4 in figure 5.2) while others are not separated by a black key (e.g. B3 and C4). We say that C4 and D4 are one **whole tone** apart (or two steps) and B3 and C4 are half tone, or one **semitone** apart, i.e. one step apart.

5.3 Major and minor scales

If we ascend (or descend) by 12 successive semitones starting from any key, we are playing the **chromatic scale**. The chromatic scale is not commonly used in popular or classical music. A more common scale starts with a C (e.g. C4) and ends with the nearest C (e.g. C5), and includes in sequence all the white keys. This scale is called the **C-major scale**. In terms of whole tones (W) and semitones (S), the C-major scale follows the pattern:

C-major scale

Note:	C	D	E	F	G	A	B	C
Separation:		W	W	S	W	W	W	S

In terms of semitone steps we have: 2, 2, 1, 2, 2, 2, 1. In the C-major scale, we have seven distinct notes (C-D-E-F-G-A-B) between the two adjacent Cs and the scale is called **heptatonic** (*hepta* means seven in Greek). The C-major scale is also a **diatonic** scale, meaning that between the two adjacent Cs, the pattern includes two semitones that are separated by at least one whole tone. There are many possibilities for heptatonic and diatonic scales. For example, starting with an A (e.g. A4) and ascending (or descending) to the nearest A (e.g. A5) and playing in sequence all the white keys, we have the **A-natural minor** scale. In terms of whole tones (W) and semitones (S), the A-natural minor scale follows the pattern:

A-natural minor scale

Note:	A	B	C	D	E	F	G	A
Separation:		W	S	W	W	S	W	W

In terms of semitone steps we have: 2, 1, 2, 2, 1, 2, 2. Comparing the two scales above, we see that they both have the *same keys*, but the position of the two semitones relative to the starting note are different, and as a result, the two scales sound very different. Generally, minor scales[2] may sound 'sad'; for example, the *funeral march* in the third movement of Mahler's Symphony N° 1. Major scales sound 'happy', like the nursery rhyme Brother John (Frère Jacques). It is interesting to note that Mahler's funeral march is a conversion of the nursery rhyme to a minor scale!

---

[2] See brief discussion of harmonic and melodic minors at the end of the chapter.

In an oversimplified picture, we can view a musical tune as a timed sequence of steps of various sizes. For example, *Twinkle Twinkle Little Star* in the C-major scale would look like:

Note:	C	C	G	G	A	A	G
Lyric:	twinkle		twinkle		little		star
Separation in semitones:		ascend 7		ascend 2		descend 2	

We can get a different tune, also in the C-major scale, using a different sequence of steps:

Note:	C	C	C	G	A	A	G
Lyric:	Old	Mac	Donald		had	a	farm
Separation in semitones:			descend 5		ascend 2		descend 2

Depending on the range of tones that a singer's voice can produce, it may be necessary to start a song from a different note. For example, *Twinkle Twinkle Little Star* can be played as follows:

Note:	F	F	C	C	D	D	C
Lyric:	twinkle		twinkle		little		star
Separation in semitones:		ascend 7		ascend 2		descend 2	

Here we **transposed** the song from the C-major scale to the F-major scale. In terms of whole tones (W) and semitones (S), the F-major scale follows the pattern:

F-major scale

Note:	F		G		A		B♭		C		C		E		F
Separation:		W		W		S		W		W		W		S	

Note that the separation is the same as for the C-major scale, i.e. in terms of semitone steps we have again: 2, 2, 1, 2, 2, 2, 1. As will be discussed in the next section, in the modern system of tuning, the semitone steps are all the same. The equivalence of steps allows us to make major scales (and minors scales) starting from any note. In other words, we can have 12 major scales, and 12 natural minor scales.

Pattern of semitone steps for all the major scales:	2, 2, 1, 2, 2, 2, 1
Pattern of semitone steps for all the natural minor scales:	2, 1, 2, 2, 1, 2, 2

Each scale is named after the first note in the scale, which is referred to as the **tonic** of the scale. Figure 5.3 shows three major and three minor scales. The tonics here are

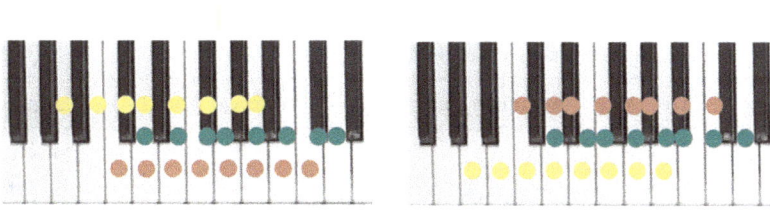

Figure 5.3. Three major and three natural minor scales: A (yellow dots), C (red dots), and D (green dots).

Table 5.1. Frequency difference and frequency ratio for two tones separated by one semitone.

Note	Frequency (Hz)	Note	Frequency (Hz)	Difference F−E (Hz)	Ratio F/E
E4	329.63	**F4**	349.23	19.60	1.0595
E5	659.26	**F5**	698.46	39.20	1.0595

A (yellow dots), C (red dots), and D (green dots). Using the patterns of semitone steps, the reader could construct the major and minor scales for any tonic.

5.4 Frequency relations and intervals

Table 5.1 lists a set of four tone frequencies used in standard piano tuning. The fifth column of the table shows that the difference between the frequencies of F4 and E4 is not the same as the difference between F5 and E5. Therefore, the difference between frequencies is not a consistent measure for specifying the frequency spacing of the tones. Looking at the sixth column, we see that the frequency ratio of F4 and E4 is the same as the frequency ratio of F5 and E5. Therefore, the frequency ratio between tones is suitable to specify the frequency spacing of the tones.

The **interval** between two tones is the ratio of the frequencies of the two tones. For the example in table 5.1, the interval between F4 and E4 is 1.0595, and is exactly equal to the interval between F5 and E5. In fact, it is 1.0595 for any consecutive Es and Fs on the keyboard. Therefore, if we know the frequency of E2, then multiplying that frequency times 1.0595 gives us the frequency of F2. More generally, if we know the frequency of any tone (black or white key), then multiplying that frequency times 1.0595 gives the frequency of the next note (in the ascending direction) on the keyboard, which may be a black or a white key. This relation between tone frequencies is the basis of the modern tuning system, which is discussed next.

5.5 The equal temperament scale

An ordinary piano is tuned so that the interval between any two adjacent notes (including black keys) is the same. This tuning is called the **equal temperament tuning** or **equal temperament scale**. Note that E and F are adjacent (there are no white or black keys between them) and according to table 5.1 the interval between them is 1.0595, which is the interval of a semitone. In the equal temperament tuning, the interval of all semitones must be the same. It follows then that all semitone intervals in the equal temperament scale are equal to 1.0595. Using the values in table 5.1, a

Table 5.2. Interval and frequency relations for the C-major scale.

Musical note	Freq. (Hz)	Semitone steps from C4	Interval from C4	Name of interval
C4	261.63		1.000	Unison
D4	293.66	2	1.122	Major second
E4	329.63	4	1.260	Major third
F4	349.23	5	1.335	Fourth
G4	392.00	7	1.498	Fifth
A4	440	9	1.682	Major sixth
B4	493.88	11	1.888	Major seventh
C5	523.25	12	2.000	Octave

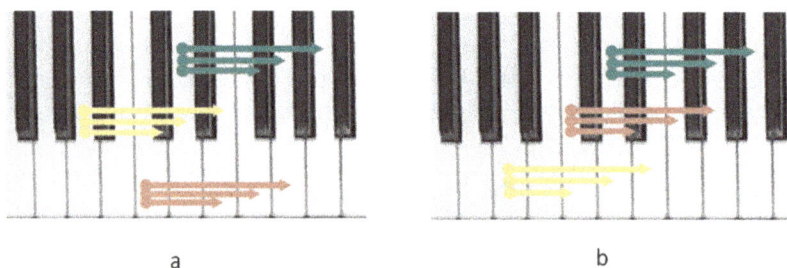

a   b

Figure 5.4. (a) Major thirds, the fourths, and fifths for A (yellow arrows), C (red arrows), and D (green arrows) tonics. (b) Natural minor thirds, the fourths, and fifths for the same tonics.

simple calculation shows that the frequency ratio (i.e. the interval) between E5 and E4 is 2, and that the frequency ratio between F5 and F4 is 2 also. The same is true for all consecutive Cs or Ds or As, etc. So if we know the frequency of E3, multiplying that frequency times 2 gives us the frequency of E4. Frequency ratio = 2 means that the frequency of E5 is twice the frequency of E4, and defines a very important interval, namely the **octave**. Since we know the interval between semitones all that we need to build a scale in terms of frequency is to choose the frequency of one key. The common choice is to assign to **A4 the frequency of 440 Hz**.

Table 5.2 lists the frequencies of the white keys in one octave in the equal temperament tuning, ascending from the middle C. As noted earlier, this is the C major scale. The fourth column lists the intervals from C4, i.e. the frequency of the note in the second column, divided by the frequency of C4. The intervals have proper names as well, which are listed in the fifth column. For example, if we ascend from C4 by two semitones, we have the major second of the C-major scale, which is D4. If we ascend another two semitones (i.e. a total of four semitones from C4) we have the major third of the C-major scale, which is E4, and so on. The proper names in the fifth column arrange the notes in a hierarchy with respect to the tonic of the scale, which in the case of table 5.2 is C4. Similar intervals are defined for all major and minor scales.

Figure 5.4(a) shows the major thirds, fourths, and fifths for three tonics; namely A (yellow arrows), C (red arrows), and D (green arrows). Figure 5.4(b) shows the minor thirds, the fourths, and fifths for three tonics; namely A (yellow arrows),

C (red arrows), and D (green arrows). Note that for a given tonic, the fourths and fifths are the same for the major and the natural minor scales.

If the two tones involved in an interval are played simultaneously, we speak of a vertical or harmonic interval. If played in sequence, we have a horizontal or melodic interval. In the same way, three or more notes form a **chord**, which can be played vertically or horizontally. In what follows we will refer to vertical intervals.

5.6 Consonance and dissonance

From ancient times, it was known that some combinations of tones sounded good together, while others did not. While we can only guess how the music of different cultures sounded, we can get a pretty good idea of the intervals used, for example, by analyzing the size and distances between holes in ancient flutes (see section 11.3). Around 500 BC, Pythagoras discovered the mathematical relation between some commonly used intervals. These intervals were known by many cultures and for many centuries before Pythagoras. The measurement of frequency had to wait for another 2000 years, but relying on the perceived pitch Pythagoras was able to establish the pitch *ratios*, i.e. the intervals for pleasant sounding intervals.

For example, he established that a pitch ratio of 3:2 sounds very pleasant. This interval is known as a **perfect fifth**. The pitch ratio of 4:3 is also pleasant and this interval is known as a **perfect fourth**. Note that in both cases, as well as other intervals, for example, the octave (ratio 2:1) or the **major third** (ratio 5:4), the frequency relation is a ratio of two **small integer numbers**. Over the centuries, several systems of tuning, i.e. ways of assigning frequencies to each tone in an octave, were developed starting with ratios of small numbers. Two such scales are discussed at the end of the chapter.

In ancient times the reason why some combinations were pleasant sounding, or **consonant**, when played together simultaneously, was attributed to some more or less mystical properties of numbers. In the past two centuries, scientific theories were advanced to explain **consonance**, and its opposite, **dissonance**. As discussed in sections 3.4 and 3.5, the tones produced by a musical instrument contain many frequencies, the fundamental and the overtones. According to one explanation, the presence of beats (see section 1.4) makes a combination of two notes dissonant. So as long as the frequencies (fundamental and overtones) of each tone are not close enough to produce perceptible beats, the interval is consonant.

Example: In the case of the perfect fifth, the two tones have a frequency ratio of 3:2. If we start with A2 (frequency = 110 Hz) the frequency of the perfect fifth is $(3/2) \times 110 = 165$ Hz which is an E3. Listed in table 5.3 are the frequencies of the first eight harmonic overtones. We note that the frequencies of some of the harmonics

Table 5.3. Frequencies of the first eight harmonic overtones of A2 and E3.

Harmonic	1	2	3	4	5	6	7	8	9
Freq. (Hz)	110	220	330	440	550	660	770	880	990
Freq. (Hz)	165	330	495	660	825	990	1155	1320	1485

Table 5.4. Overtone frequencies of A2, G2♯, and C4♯.

Harmonic	1	2	3	4	5	6	7	8	9
Freq. (Hz)	104	209	313	418	522	626	731	835	940
Freq. (Hz)	110	220	330	440	550	660	770	880	990
Freq. (Hz)	277	554	831	1108					

of the two tones match exactly (highlighted in the table). We also note that the values of the remaining frequencies are too far apart (by 55 Hz or more) to produce perceptible beats. Therefore, this pair of tones is consonant.

The situation is different if we consider, for example, two tones differing by one semitone, or two tones whose overtones may be too close. Table 5.4 lists the harmonics of 104 Hz, 110 Hz, and 277 Hz, which correspond to A2, G2♯, and C4♯, respectively.

Note that all the frequencies are close together, and there are no matching frequencies between harmonics. As the difference between frequencies is small (6 Hz in the second column), perceptible beats will occur. Therefore, G2♯ and A2 played simultaneously is dissonant. Note also that the fifth harmonic of A2 is very close to the second harmonic of C4♯, therefore when the two notes are played together some roughness may occur.

The concept of the critical band, introduced in section 4.11, provides a different basis to explain dissonance. According to this theory, two notes that differ in frequency by less than *half of the width of the critical band* at that range of frequency will have some dissonance, which is usually referred to as **roughness**. The dissonance is more noticeable the smaller the difference between the two frequencies. For example, for the frequencies listed in table 5.3, the frequency difference is 55 Hz. The width of the critical band in the range of 100–200 Hz is about 90 Hz. Half the width is 45 Hz, which is smaller than the 55 Hz minimum separation between the frequencies listed in table 5.3. Therefore, the two tones in table 5.3 are consonant. On the other hand, the first two frequencies listed in table 5.4 differ by 6 Hz (11 Hz in the third column, 17 Hz in the fourth column, etc) and the width of the critical band is 90 Hz. The frequency difference is considerably smaller than half the bandwidth and the two notes are dissonant if played simultaneously.

5.7 From dissonance to consonance

From the intervals in equal tempered tuning listed in table 5.2, we see that the fifth and the fourth (1.498 and 1.335, respectively) are very close to the perfect fifth (3:2 = 1.5) and perfect fourth (4:3 = 1.333). The equal tempered interval for the major third (1.260) shows a larger deviation from the simple ratio 5:4 = 1.250, therefore, the major third in the equal temperament tuning was considered dissonant a couple of centuries ago. It is reasonable to assume that music would more likely sound harmonious if it includes consonant intervals, which evoke relaxation. This was the common understanding in the polyphonic music of the renaissance, with many voices singing different tones simultaneously. Dissonant intervals are considered to evoke tension (among other things) and as such are unstable and require 'resolution'. As it happened over the centuries, intervals that were considered dissonant found their way into major works. Therefore, a dissonant vertical interval is not necessarily unusable, and composers use the succession of consonance and dissonance to produce emotionally rich and powerful music.

5.8 Other scales

Our previous discussion was based on the heptatonic scale (e.g. ABCDEFGA), which is the most common. A related scale is the **pentatonic** scale, which uses five tones[3]. Pentatonic scales of various kinds are found in jazz, country music, and Native American songs, to name a few. One can very simply play a major pentatonic scale, by ascending on the black keys of the piano, starting from F♯. A C-major scale would be CDEGAC. In this particular case, the interval from E to G is three semitones, as is the interval from A to C. There are no single semitone intervals. The particular scale derives from the heptatonic C-major by omitting the fourth (F) and seventh (B) tones. A pentatonic scale in A-minor can be derived from the corresponding heptatonic scale by omitting the second (B) and sixth (F) tone, i.e. use ACDEGA. Again, in this particular pentatonic minor there are two three-semitone intervals (from A to C and from E to G) and there are no single semitone intervals.

A piano can produce only a certain set of frequencies in steps of one semitone. For example, in the standard tuning (see table 5.2) the E4 will produce 329.63 Hz and the F4 will produce 349.23 Hz. As there is no key in between E and F, we cannot readily produce an intermediate frequency, say 339 Hz. This is not that case for all instruments. A violin, for example (section 11.4), can produce all the frequencies intermediate between E4 and F4. As such, the violin can play the perfect fourths and fifths discussed above. In addition, the violin can play steps of half semitones, the so-called **quartertones**. Arabic music, Persian music, North African, and others use quartertone scales. Such scales have 24, rather than 12 steps within an octave. The larger number of steps gives a larger number of combinations for constructing the equivalent of major and minor scales from each tonic. For example, the traditional Arabic system has over 40 such scales known as **maqamat** (plural of **maqam**), and more variations in North African, Persian, and Turkish music.

5.9 Older temperaments

Our discussion in this chapter is based on the equal temperament scale. In this section, we describe two scales that played a significant role in the development of Western music, namely the **Pythagorean** and the **just scales**.

One of the earlier scales, named after Pythagoras, is constructed using perfect fifths (frequency ratio 3:2) and octaves (frequency ratio 2:1). We refer to the

---

[3] *Pente* means five in Greek.

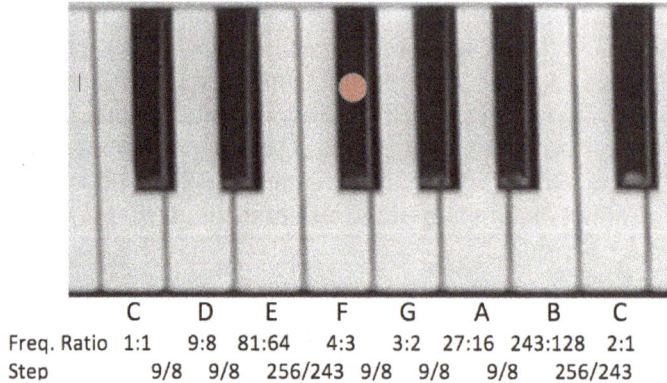

Figure 5.5. Pythagorean intervals for the C-major scale.

frequency of each tone by the ratio of its frequency to the frequency of the starting tone. For example, if we start with C3, then the frequency ratio for the perfect fifth is 3:2, and the frequency ratio for the octave, i.e. C4, is 2:1. It is important to keep in mind that ascending by one fifth from any note means multiplying the frequency ratio of that note times 3:2. Descending by one perfect fifth from any note means multiplying the frequency ratio of that note times 2:3. The same applies for ascending or descending by any interval. For example, ascending by a perfect fourth means multiplying the frequency of the starting note times 4:3, and descending means multiplying the frequency of the starting note times 3:4.

Figure 5.5 shows the Pythagorean intervals for the C-major scale[4]. Ascending or descending the white keys starting from C we have the so-called **C-major scale**. We see that any pair of white keys in the C-major scale separated by seven steps (counting both white and black keys) is a perfect fifth. We can verify this fact by dividing frequency ratios. For example, dividing the frequency ratio of A (i.e. 27:16) by the frequency ratio of D (i.e. 9:8), we find $(27:16)/(9:8) = 3:2 = 1.5$. One can use a calculator to verify the ratios for the other two pairs (B&E and D&A). One thing that we note is the absence of the major third ($5:4 = 1.25$) that should correspond to E. Instead we see that E corresponds to $81:64 = 1.27$. Also, recall that the intervals that sound pleasing should correspond to ratios of two small integer numbers, and 81:64 is not a ratio of small integers.

The frequency ratios for successive white keys are listed at the bottom of figure 5.5. Thus C&D, D&E, F&G, G&A, and A&B are all separated by two keys or two semitone steps (i.e. one whole tone) and from the bottom row we see that the full tone interval here is $9:8 = 1.125$. E&F and B&C are one step apart, or one semitone, therefore the semitone interval here is $256/243 = 1.053$. At this point we see one problem with the Pythagorean scale, namely a two-semitone interval (i.e. $1.053 \times 1.053 = 1.110$) is not equal the full tone interval of 1.125. This can cause problems with assigning the frequencies for the black keys. For example, we can look at the frequency of F♯ (or G♭) indicated in figure 5.5 by a red dot. To do so, we can ascend

---

[4] The procedure used to construct the Pythagorean scale is outlined at the end of this chapter.

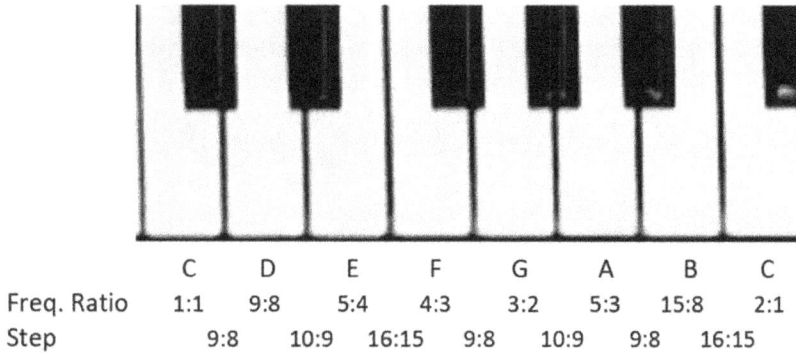

	C	D	E	F	G	A	B	C
Freq. Ratio	1:1	9:8	5:4	4:3	3:2	5:3	15:8	2:1
Step		9:8	10:9	16:15	9:8	10:9	9:8	16:15

Figure 5.6. Intervals for the C-major scale in the just tuning.

from F (4:3) by one semitone (i.e. multiply 4:3 times 256:243 = 1024:729 = 1.404) or descend from G (3:2) by one semitone (i.e. multiply 3:2 times 243:256 = 729:512 = 1.424). The ratios are not the same, therefore we have an ambiguity is assigning a frequency to F♯ (G♭). The conclusion is that the semitone steps in the Pythagorean scale are not the same across the C-major scale, and this can cause problems with assigning some frequencies. As long as we stay on the C-major scale there is no problem. The problems arise when we try to transpose from the C-major to other tonics.

The **just scale** avoids the difficulty of the major third, discussed above in the Pythagorean scale, by constructing a scale that uses both the perfect fifth and the major third, as follows. A **triad** comprises three tones. A **major triad** combines a major third and a perfect fifth. For example, starting from C, a major third from C is E, and a major fifth from C is G. The triad C-E-G is called the **C-major triad**. Figure 5.6 shows the intervals for the C-major scale[5]. Here we see that the major third, and the perfect fourth, and fifth have the right ratios, namely 5:4, 4:3, and 3:2, respectively. Therefore, the scale includes very consonant intervals. The problem with the just scale is in the whole tone intervals. We note that C&D, D&E, F&G, G&A, and A&B are all separated by a whole tone, but the step between C&D, F&G, and A&B is 9:8 (=1.125) while the step between D&E, and between G&A is 10:9 (=1.111). In other words, the whole tone intervals are not the same across the scale. Also, E&F and B&C are separated by one semitone step and the interval is 16:15 = 1.0667. Now a whole tone step is two semitone steps, therefore the ratio should be 1.0667 × 1.0667 = 1.1378. This ratio neither matches 9:8 (1.125) nor 10:9 (=1.111). This mismatch causes problems with transposing, as in the Pythagorean scale.

5.10 Noise

In everyday conversation we use the word *noise* to indicate an unpleasant or loud sound. In science noise can indicate unwanted signals in sound equipment (see section 8.8) or a mixture of all frequencies. **White noise** is a random mix

---

[5] The procedure used to construct the just scale is outlined at the end of this chapter.

of frequencies of equal intensity. Blowing the sound 'fff' would be a close approximation to white noise. **Pink noise** is a random mix of frequencies with equal intensity in each octave. Noise is not necessarily a useless thing. For example, in flutes we essentially start by blowing all frequencies into a pipe. As described in section 11.3, the pipe channels the energy of the noise into musical notes.

5.11 Further discussion

Structure of scales

It is important to keep in mind that taking a step in a musical scale means multiplying, not adding. The frequency ratio of the starting note is the unison, i.e. 1:1 = 1. When forming the frequency ratio of a note, we divide the frequency of the note in question by the frequency of the starting note. Ascending by one perfect fifth from any note means multiplying the frequency ratio of that note times 3:2 (= 1.5). Descending by one perfect fifth from any note means multiplying the frequency ratio of that note times the inverse of the interval, i.e. multiplying times 2:3 (=0.667). The same applies for ascending or descending by any interval. For example, ascending by a perfect fourth means multiplying the frequency ratio of that note times 4:3, and descending means multiplying times 3:4. Also, ascending by one octave from any note means multiplying the frequency ratio of that note times 2. Descending by one octave from any note means multiplying the frequency ratio of that note times ½.

a. *Structure of the equal temperament scale*
   In the equal temperament scale, we have 12 equal semitone intervals in one octave. The frequency doubles within one octave. Therefore, the semitone interval multiplied times itself 12 times should give 2. In other words, the semitone interval must be the twelfth root of 2. Thus the **semitone interval in equal temperament** $\sqrt[12]{2} = 1.0595$ in agreement with the value listed in the sixth column of table 5.1.

b. *Structure of the Pythagorean scale*
   The Pythagorean scale is constructed using two intervals, namely the octave and perfect fifths. Here we will use the symbols P5 for the perfect fifth and P8 for the octave.

Suppose that we start with C4 (1:1, unison).
- Ascending by P5 from C4 gives us G4, the perfect fifth.
- Ascending by P8 from C4 gives us C5 (2:1, octave).
- Descending by P5 from C5 gives us (2:1) × (2:3) = 4:3 which is the perfect fourth F4.
- Ascending by P5 from G4 gives us (3:2) × (3:2) = 9:4. This number is larger than 2, i.e. a frequency ratio higher than C5, therefore we need to descend by one octave. Descending by P8 from 9:4 takes us to (9:4) × (1:2) = 9:8 which is the major second D4. The steps in this sequence are shown in the figure.

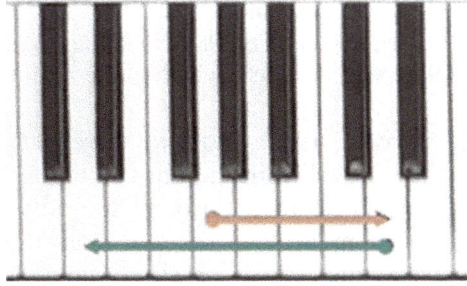

- Ascending by P5 from D4 gives us $(9:8) \times (3:2) = 27:16$ which is the major sixth **A4**.
- Descending by P8 from A4 gives us $(27:16) \times (1:2) = 27:32$. From here we ascend by P5 which gives us $(27:32) \times (3:2) = 81:64$ which is the major third **E4**. The steps in this sequence are shown in the figure.

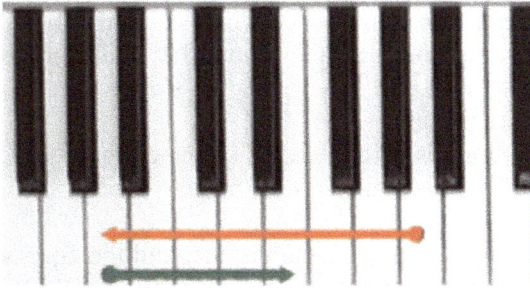

- Ascending by P5 from E4 gives us $(81:64) \times (3:2) = 243:128$ which is **B4**.

With the above steps we created all the white keys in the C4–C5 octave (highlighted in yellow), and the frequency ratios in the Pythagorean tuning as listed in figure 5.5.

c. *Structure of the just scale*

The just scale is based on triads. We will go through the steps of constructing the C-major scale, starting from C4. The major triad already has **C4** (1:1), **E4** (5:4), and G4 (3:2), and of course we have the octave, **C5** (2:1).

Ascending one major triad from any note in the above triad, will give us the note itself and two more notes. The frequency ratios of the new triad are:

- (ratio of new starting note) $\times$ (1:1), i.e. the note itself and
- (ratio of new starting note) $\times$ (5:4),
- (ratio of new starting note) $\times$ (3:2).

For example, we can form a new triad starting from G4. Here the ratio of the new starting note is 3:2 and following the steps above the frequency ratios of the new triad are:

- $(3:2) \times (1:1) = 3:2$, i.e. the G4 the start of the new triad and
- $(3:2) \times (5:4) = 15:8$, i.e. **B4** the major seventh of the C-major scale,

- $(3:2) \times (3:2) = 9:4$ which is a number higher than 2, i.e. a frequency ratio higher than C5. We can descend[6] by P8, which means multiplying times $(1:2)$, and the result is $(9:4) \times (1:2) = 9:8$, which is the major second **D4**.

Descending one major triad from any note will give us the note itself and two more notes. Note that the interval from the third to the fifth is $(3:2)/(5:4) = 6:5$, therefore to descend from a fifth to a third we multiply times the inverse of 6:5, i.e. multiply times 5:6. Similarly, to descend from the fifth to the first we multiply times the inverse of the fifth, i.e. multiply times 2:3. If we form a new triad by descending by one major triad from any note, then this note is the *fifth* of the new triad. Thus, descending by one triad from any note we follow the procedure:

- (ratio of new fifth) $\times$ (1:1) this is the fifth of the new triad,
- (ratio of new fifth) $\times$ (5:6) this is the third of the new triad,
- (ratio of new fifth) $\times$ (2:3) this is the first of the new triad.

For example, we can form a new triad where C4 is the (new) *fifth*. Here we are keeping the same starting note, and the ratio is 1:1. Following the steps above the frequencies of the new triad are:

- $(1:1) \times (1:1)$, i.e. C4, and
- $(1:1) \times (5:6) = 5:6$ this is the third of the new triad,
- $(1:1) \times (2:3) = 2:3$ this is the first of the new triad.

Note that the last two ratios are less that one, and the notes are lower than C4. Ascending by P8 from each, we have:

- $(5:6) \times (2:1) = 10:6 = 5:3$, which is the major sixth **A4** of the C-major scale, and
- $(2:3) \times (2:1) = 4:3$ which is the major fourth **F4**.

With the above steps we created the C-major scale (highlighted in yellow), and the frequency ratios in the just tuning as listed in figure 5.6.

Other minor scales

In section 5.3, we discussed the natural minor scale. Other commonly used minor scales are the **harmonic minor** and the **melodic minor**.

The pattern of semitone steps for all the natural minor scales:	**2, 1, 2, 2, 1, 2, 2**
The pattern of semitone steps for all the harmonic minor scales:	**2, 1, 2, 2, 1, 3, 1**

The patterns above are the same for ascending and descending. Note that the harmonic minor has three semitone steps, and one step of three semitones. Therefore, the harmonic minor does not follow the rule for diatonic scales, as defined in section 5.3. The pattern for the melodic minor has two semitones

---

[6] Here again we us the abbreviation P8 for the octave.

separated by at least one whole tone, but the pattern is different for ascending and descending.

Ascending pattern of semitone steps for all the melodic minor scales:	**2, 1, 2, 2, 2, 2, 1**
Ascending pattern of semitone steps for all the melodic minor scales:	**2, 1, 2, 2, 1, 2, 2**

The descending pattern listed above must be read from right to left. It is interesting to note that the first four intervals for all the minor scales listed above are the same. Having different ascending and descending patterns is not unique to the melodic minor. Many scales used in Arabic, Persian, and other Eastern and North African music have different ascending and descending patterns.

5.12 Questions

1) (a) In what way does the heptatonic scale differ from the chromatic scale?
    (b) What is the major advantage of the equal temperament scale compared to the Pythagorean and the just scales?

2) (a) Which major scale(s) can be played using only the white keys of the keyboard?
    (b) For each scale, name the notes corresponding to the major third, the fourth and the fifth.
    (c) Which minor scale(s) can be played using only the white keys on the keyboard?
    (d) For each scale, name the notes corresponding to the minor third, the fourth and the fifth.

3) Using 440 Hz for A4 is a recent standard. In the Baroque era, A4 = 415 Hz was commonly used.
    (a) Is the present standard sharper or flatter than the baroque standard?
    (b) Find the ratio of the frequencies of A4 (present) and A4 (baroque).
    (c) What is the value of one semitone interval in the equal temperament scale?
    (d) Recalling that the interval between two tones is given by the ratio of their frequencies, how far apart is the Baroque A4 from the modern A4?

4) Use figure 5.1 and the patterns of semitone steps in section 5.3 to construct:
    (a) the A-major scale
    (b) the natural C-minor scale

Chapter 6

Standing waves and resonance

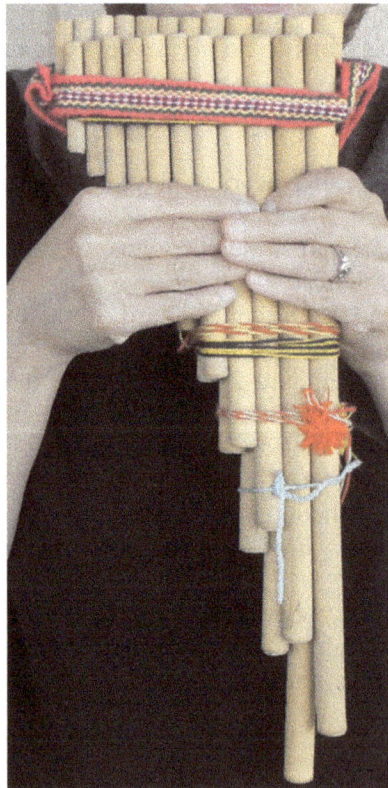

6.1 Introduction

So far, we have described waves that propagate in open space. Propagating waves travel at a speed determined by the medium. For sound waves in air, there are no restrictions on the frequency of the wave, i.e. all audible frequencies propagate in open air. A different type of wave motion occurs when the wave is confined; for example, the air inside a glass bottle. When blowing air across the mouth of an empty bottle, a characteristic tone is heard. The frequency of the tone produced depends on the size of the bottle. A question that comes to mind is: why do we hear just one pitch? The tone heard is not produced by vibrations in the glass, but by vibrations of the air inside the bottle. Blowing air in a half-filled bottle produces a sound of higher pitch. This observation suggests that the frequency of the tone becomes higher as the height of the air column in the bottle gets shorter. This suggests a relation between the tone pitch and the amount of air in the bottle. By choosing a set of bottles of the right size and shape, one can create a kitchen version of a pan flute. A pan flute, shown in the opening figure, is essentially a collection of pipes open at one end and closed at the other end. The length of the air column inside the tube determines the tone produced by each tube. The relation between length of the air column and the frequency gives a simple way of selecting the musical tones produced by the instrument. This is the basis of constructing wind instruments, as we will see in section 11.3. A similar relation between length and frequency occurs in strings, and is the basis of string instruments. The vibrational motion excited in pipes and strings is an example of **standing waves**. In this chapter, we discuss the relation between the geometric dimensions and the frequency of the standing waves produced in pipes and strings. The vibrations of plates and membranes will be discussed in section 11.5. We will start by discussing standing waves in strings.

6.2 Vibrational modes in a string

Consider a string that is clamped at both ends; for example, a guitar string. The two endpoints of the string are clamped, so the displacement is always zero at the endpoints. The string can only sustain vibrations that are compatible with the requirement that the displacement at the two endpoints remains zero at all times. We will call this condition the **boundary condition**. Plucking a string excites a vibration, in which each point of the string **oscillates** about the equilibrium position, except for the two endpoints, which are clamped. In this oscillation, the restoring force is the tension of the string. What is different here is that the 'wave' is confined between the two endpoints. This is a **standing wave**, as opposed to the traveling waves discussed in chapter 2. We can find the allowed vibrations by looking at the wavelength[1].

If the string is flexible, and is plucked gently, the vibration pattern is sinusoidal, as shown in figure 6.1. The amplitude of the oscillation is different at each point of the string. At the endpoints the amplitude is zero, because the endpoints are clamped. In the middle of the string the amplitude is at maximum. Figure 6.1 shows 'snapshots' of the vibrating string. The sinusoidal pattern drawn corresponds to half a cycle of a sinusoidal, or 1/2 of a wavelength. Therefore, if half the wavelength of the vibration equals the length of the string, the vibration is allowed. In other words, vibration is sustained if the wavelength of the vibration equals twice the length of the string.

---

[1] A mathematical derivation is given in appendix B.4.

Length of String

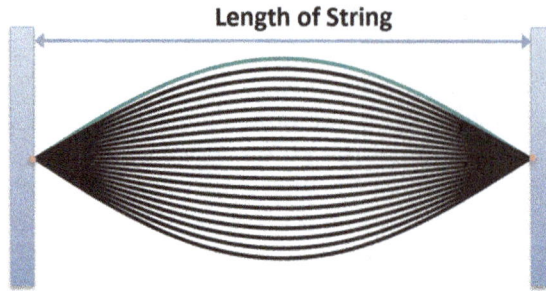

Figure 6.1. Snapshots of a standing wave pattern in a guitar string.

Length of String

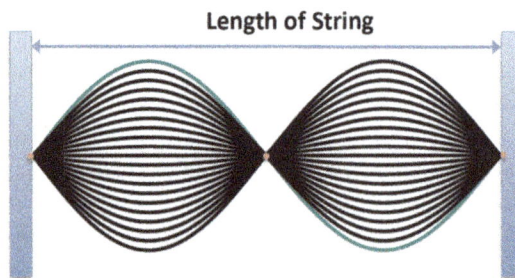

Figure 6.2. A standing wave pattern that is compatible with the boundary condition.

Length of String

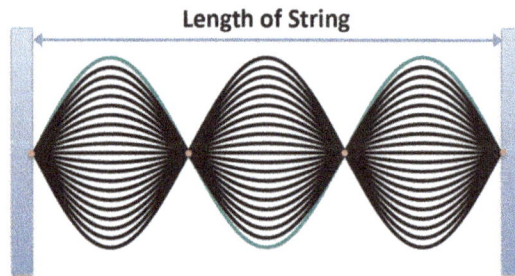

Figure 6.3. A standing wave of three half-cycles.

The pattern shown is not the only sinusoidal that is compatible with the boundary condition. As it turns out, any sinusoidal that satisfies the boundary condition represents a vibration that can be sustained by the string. Figure 6.2 shows another sinusoidal compatible with the boundary condition, therefore, a vibration that can be sustained by the string. In this case, one complete cycle (i.e. two half-wavelengths) of the sinusoidal equals the length of the string.

Figure 6.3 shows yet another waveform compatible with the boundary condition. As indicated by the green line in the figure, the sinusoidal completes one and one half cycles, or three half-cycles. Therefore, the length of the string equals 3/2 wavelengths; in other words, the wavelength is 2/3 of the length of the string.

Comparing figures 6.1–6.3, we see that the wave in figure 6.1 has the longest wavelength of all. This is the **fundamental mode** of the vibration. The next longest is the mode shown in figure 6.2. This is the second vibrational **mode, or second harmonic.** In figure 6.3, the wavelength is shorter than the previous two. This is the third mode or third harmonic. It follows that the vibration is allowed as long as we can fit an integer number of half-wavelengths into the length of the string. In other words, a vibration is sustained whenever:

$$n \times (\text{wavelength}/2) = \text{length of string}$$

where n can be 1, 2, 3, 4, etc. For $n = 1$ we have the fundamental mode, $n = 2$ is the second mode, and so on. Comparing the wavelengths from figures 6.1–6.3, we see that there is a direct relation between the wavelength and the order of the harmonic:
- The wavelength of the second harmonic is 1/2 the wavelength of the fundamental.
- The wavelength of the third harmonic is 1/3 the wavelength of the fundamental.

By adding half-wavelengths as done above, one can see a trend in how the wavelengths relate to the fundamental:
- The wavelength of the fourth harmonic is 1/4 the wavelength of the fundamental.
- The wavelength of the nth harmonic is $1/n$ the wavelength of the fundamental, where n is any positive integer (1, 2, 3, and so on).

Example: As an application we consider the vibration modes of the fifth string (fifth from bottom) of a guitar. In standard tuning, the open fifth string produces a frequency of 110 Hz. When the string is open, the two endpoints[2] are the **saddle** (on the body of the guitar) and the **nut** (at the top of the fretboard). Typically, the distance between saddle and nut is 0.65 m.

As discussed above, the length of the string is equal to 1/2 the wavelength of the fundamental mode, i.e. the wavelength of the fundamental mode is $2 \times 0.65 = 1.3$ m. According to the discussion above:
- The wavelength of the second vibrational mode equals $1.3/2 = 0.65$ m.
- The wavelength of the third vibrational mode equals $1.3/3 = 0.433$ m, and so on.

Recall from section 1.2 that the frequency is related to the wavelength and the speed of the wave by the equation

$$\text{speed} = (\text{frequency}) \times (\text{wavelength}).$$

---

[2] See section 11.4 for the parts of a guitar.

Table 6.1. Wavelengths and frequencies of the vibrational modes of the fifth string of a guitar.

Mode	Wavelength (m)	Frequency (Hz)
$n = 1$	1.30	110
$n = 2$	0.65	220
$n = 3$	0.433	330
$n = 4$	0.33	440

If the speed of sound in the string is not changing, then as the wavelength decreases, the frequency must increase to keep the product unchanged. Hence it follows that if the wavelength decreases by a factor of 2 (or 3, or 4, etc) the frequency must increase accordingly by a factor of 2 (or 3, or 4, etc). Therefore, we conclude that the frequencies of the vibrational modes must be
   • The frequency of the fundamental is 110 Hz.
   • The frequency of the second vibration mode equals $2 \times 110 = 220$ Hz.
   • The frequency of the third vibration mode equals $3 \times 110 = 330$ Hz, and so on.

   Table 6.1 lists the wavelengths and frequencies of the first four modes for the fifth string of a guitar. The length is assumed to be 0.65 m.
   The frequency of the modes is an integer multiple of the fundamental, i.e. we have the following the pattern for the mode frequencies: 1, 2, 3, 4, 5 times the fundamental, and so on. This pattern is the **harmonic series.**
   Recall that the product of the frequency times the wavelength for each mode gives the speed of the wave in the string:

Fundamental mode:	110 Hz $\times$ 1.3 m = 143 m s^{-1}
2nd vibrational mode:	220 Hz $\times$ 0.65 m = 143 m s^{-1}
3rd vibrational mode:	330 Hz $\times$ 0.433 m = 143 m s^{-1} and so on.

   Note that the speed is the same for all vibrational modes. The speed of the wave in the string depends on the magnitude of the restoring force (i.e. the tension of the string) and the nature of the string (specifically the thickness and the material). Therefore, for a given string, the speed is the same for all modes. Also note that the speed of 143 m s^{-1} refers to the speed of the waves in the string, *not* the speed of sound in air, which is about 343 m s^{-1}.

6.3 Nodes and antinodes

From figures 6.1–6.3 we note that the fundamental mode of vibration has maximum amplitude at one point (in the middle) and zero amplitude at two points (the endpoints). In the second mode (figure 6.2) the amplitude is maximized at two points, and is zero at three points. Similarly, for the third mode (figure 6.3) the amplitude has three maxima, and four zeros. Following this trend, the fourth mode will have four maxima and five zeros, and so on. Points on the string that do not oscillate (i.e. the amplitude is zero) are referred to as **nodes**. Points where the oscillation is maximized are **antinodes**.

6.4 Simultaneously vibrating modes

From the above discussion we conclude that a guitar string can vibrate in many frequencies; in other words, we hear many pure tones at once. In the above example, besides the fundamental (110 Hz) we have all multiples of 110 Hz (220, 330, 440, etc) oscillating as well. Therefore, the tone produced by plucking a string is not a **pure tone** (i.e. a single frequency), but a **complex tone**, consisting of the many frequencies or **partials**, meaning the fundamental plus the overtones. The presence of multiple modes plays a key role in what we call quality or **timbre** of a tone.

Which harmonics oscillate and how strongly depend on *where* and *how* the string is plucked. By proper plucking, musicians can control the timbre of the tone. One simple method suggests itself by examining figure 6.2, which shows that the second harmonic has a node at the midpoint. By plucking the string at the midpoint, we force motion (i.e. an antinode) at the midpoint, which is not compatible with the second harmonic. Plucking the string at the midpoint will suppress the second harmonic. Note that all even (e.g. the fourth, sixth, etc) harmonics have an odd number of nodes, meaning that the midpoint for all even harmonics must be a node. It follows that by plucking the string at the middle, we suppress all the even harmonics.

6.5 Standing waves in pipes

Standing waves occur in pipes as well, and this is the basis of wind instruments. Here the vibration of interest is a longitudinal vibration of the air column *inside* the pipe, not the vibration of the pipe walls. As discussed in section 1.2 the wave consists of compressions and rarefactions, i.e. the pressure in the air column gets high and low *compared to the ambient* (or *atmospheric*) pressure. Here we will describe the standing waves in a pipe using the concept of nodes and antinodes introduced in section 6.3. For musical instruments we have two types of pipes that are of interest, namely: **open pipe**, which is a pipe open at both ends, and **semi-closed pipe**, which is a pipe open at one end and closed at the other.

As in the case of standing waves in a string, the important starting point is the boundary condition. For a clamped string, the two endpoints do not oscillate. Therefore, we have a node at both ends. In the case of a pipe, we have two possibilities, namely *open* or *closed* end, which impose different boundary conditions. At the open end of a pipe, the pressure must be equal to the ambient pressure at all times. Therefore, at the open end of a pipe, the pressure must remain constant and no oscillation is allowed. The result is that **at the open end of a pipe we must have a pressure node.**

At the closed end of a pipe, the air can be compressed to a maximum pressure and then bounce back to a minimum pressure. Therefore, at the closed end the oscillations of the pressure are the largest. The result is that **at the closed end of a pipe we must have a pressure antinode.**

We will use the above boundary conditions to investigate the mode frequencies of the standing waves supported by an open and a semi-closed pipe.

6.6 Standing waves in an open pipe

For the open pipe, the boundary conditions are the same as the clamped string, i.e. **the two endpoints are pressure nodes.** Therefore, using the same arguments as in section 6.2 we conclude that the open pipe will sustain standing waves if the length of the pipe is equal to an integer number of half-wavelengths.

Modes in open pipe:

$$n \times (\text{wavelength}/2) = \text{length of pipe}$$

where n can be 1, 2, 3, 4, etc. For $n = 1$ we have the fundamental mode, $n = 2$ is the second mode, etc.

Figure 6.4 shows the fundamental mode in an open pipe which is identical to figure 6.1. What is different is the nature of the oscillating quantity. In the case of the string (figure 6.1) the graphs show snapshots of the displacement of the string from the equilibrium position. Figure 6.4 shows snapshots of the variation of the air pressure in the pipe. The pressure at the two ends of the pipe is equal to the atmospheric pressure. With this understanding, one can interpret figures 6.2 and 6.3 as showing the second and third harmonics in an open pipe.

We can rearrange the quantities in the above relation to find the wavelengths of the modes in an open pipe:

Mode wavelengths in open pipe:

$$\text{wavelength} = (2/n) \times \text{length of pipe}$$

where $n = 1, 2, 3, 4$, etc. From the above relation we conclude that longer pipes produce fundamentals of longer wavelength.

Example: For an open pipe of length 1.70 m, find the wavelength of the fundamental mode.

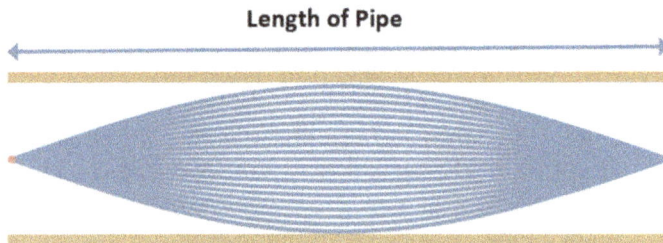

Length of Pipe

Figure 6.4. Fundamental mode in an open pipe.

Table 6.2. Wavelengths and frequencies for the first four modes in a 1.7 m long open pipe.

Mode	Wavelength (m)	Frequency (Hz)
$n = 1$	3.40	101
$n = 2$	1.70	202
$n = 3$	1.13	303
$n = 4$	0.85	404

For the fundamental mode $n = 1$, applying the mode wavelength relation we have:

$$\text{wavelength} = (2/1) \times 1.70 = 2 \times 1.70 = 3.40 \text{ m}.$$

Since frequency × wavelength equals the speed of the wave, it follows that the frequency is equal to the ratio of the speed divided by the wavelength. Here the speed of the wave is the speed of sound in air, i.e., 343 m s^{-1}. Using the mode wavelengths from the above relation, we can find the mode frequencies (in Hz) for the open pipe:

Mode frequencies in open pipe:

$$\text{frequency} = (n \times 343)/(2 \times \text{length of pipe}).$$

In the above relation, the length of the pipe is in the denominator, therefore, the longer the pipe, the lower the frequency of fundamental.

Example: For an open pipe of length 1.70 m, find the frequency of the fundamental mode.

For the fundamental mode $n = 1$, and applying the mode frequency relation we have:

$$\text{frequency} = (1 \times 343)/(2 \times 1.70) = 343/3.4 = 101 \text{ Hz}.$$

Table 6.2 lists the wavelengths and frequencies for the first four modes in an open pipe of length 1.70 m.

We note that the frequency of the modes follows the harmonic series, which was introduced in the discussion of the vibrating string.

6.7. Standing waves in a semi-closed pipe

In a semi-closed pipe, we must have a **pressure node at the open end** and a **pressure antinode at the closed end**. A half-wavelength has nodes at both ends, so it is not compatible with boundary conditions. To have an antinode at the closed end we need a quarter-wavelength.

Figure 6.5 shows the first three modes for a semi-closed pipe. For the fundamental mode, shown in figure 6.5(a), we note that the length of the pipe is equal to 1/4 wavelength. For the next higher mode, shown in figure 6.5(b), the length of the pipe

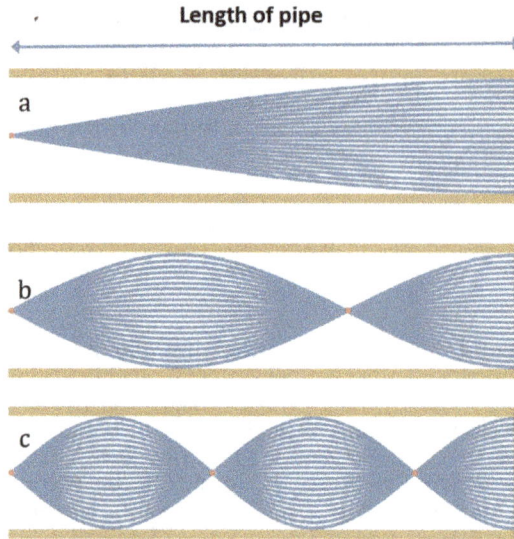

Figure 6.5. First three modes for a semi-closed pipe.

is equal to 3/4 of a wavelength. Comparing figures 6.5(a)–(c), we conclude that successive modes differ by 1/2 wavelength, or two quarter-wavelengths. Note that the fundamental fits one quarter-wavelength in the length of the pipe. Therefore, the sequence of modes in terms of *quarter-wavelengths* will be $(0 + 1), (2 + 1), (2 + 2 + 1),$ $(2 + 2 + 2 + 1),$ etc, or 1, 3, 5, 7 quarter-wavelengths, and so on.

Following this trend we can write:

Modes in semi-closed pipe:

$$n \times (\text{wavelength}/4) = \text{length of pipe}$$

where n follows the sequence of odd numbers, i.e., 1, 3, 5, 7, etc. For $n = 1$ we have the fundamental mode, $n = 3$ is the second mode, $n = 5$ is the third mode, and so on.

Rearranging the quantities in the above relation we can find the wavelengths of the modes in a semi-closed pipe:

Mode wavelengths in semi-closed pipe:

$$\text{wavelength} = (4/n) \times \text{length of pipe}$$

where $n = 1, 3, 5, 7$, etc.

Example: Find the wavelength of the fundamental mode for a semi-closed pipe of length equal 0.85 m.

For the fundamental mode $n = 1$, applying the mode wavelength relation we have:

$$\text{wavelength} = (4/1) \times 0.85 = 4 \times 0.85 = 3.40 \text{ m}.$$

Following the procedure used above in connection with open pipes, we can find the mode frequencies (in Hz) for the semi-closed pipe:

Table 6.3. Wavelengths and frequencies for the first four modes in a 0.85 m long semi-closed pipe.

Mode	Wavelength (m)	Frequency (Hz)
$n = 1$	3.40	101
$n = 3$	1.13	303
$n = 5$	0.68	504
$n = 7$	0.49	706

Mode frequencies in semi-closed pipe:

$$\text{frequency} = (n \times 343)/(4 \times \text{length of pipe}).$$

Example: For a semi-closed pipe of length 0.85 m find the frequency of the fundamental mode.

For the fundamental mode $n = 1$. Applying the mode frequency relation we have:

$$\text{frequency} = (1 \times 343)/(4 \times 0.85) = 343/3.40 = 101 \text{ Hz}.$$

Table 6.3 lists the wavelengths and frequencies for the first four modes in a semi-closed pipe of length 0.85 m.

6.8 Comparison of open and semi-closed pipes

Open and semi-closed pipes share some common properties. The fundamental frequency is higher for shorter pipes. The reverse is true for the wavelength of the fundamental: the wavelength is shorter for shorter pipes. On the other hand, there are two essential differences between the two types of pipes, which can be seen by comparing the values listed in tables 6.2 and 6.3. The first difference is that the semi-closed pipe produces only odd multiples of the fundamental frequency, while the open pipe produces both odd and even multiples of the fundamental frequency. Therefore, the timbre of the two types of pipes is different. Second, we note that the semi-closed pipe (0.85 m long in the above example) produces the same fundamental frequency as an open pipe twice as long (1.70 m in the example). Therefore, semi-closed wind instruments (see section 11.3) can be shorter than open instruments, and produce the same tonal range.

6.9 Standing waves in a pipe closed at both ends

Finally, the mode frequencies for a pipe closed at both ends (closed pipe) are the same as the open pipe:

Mode frequencies in closed pipe:

$$\text{frequency} = (n \times 343)/(2 \times \text{length of pipe})$$

where n takes the values 1, 2, 3, etc.

Pipes closed at both ends are not relevant to musical instruments, since the sound cannot escape from the closed pipe. However, the pipe closed at both ends is a useful model for standing waves in rooms, as will be discussed in section 9.11.

6.10 Standing waves in rods and tubes

Of interest to percussion instruments are standing waves that form in solid rods and in tubes. Solids can transmit both longitudinal *and* transverse waves, and can sustain both longitudinal and transverse standing waves. The behavior of rods and tubes is essentially the same, and we will use *rods* to refer to both. The frequency of the fundamental mode depends on the length of the rod. The frequency of the fundamental is lower for longer rods. This trend is similar to what we found for strings and air columns in pipes. The frequency of the fundamental mode depends also on the stiffness of the material. For example, the fundamental frequencies of wooden or bamboo rods are lower than those of steel or aluminum. The stiffness of the material acts as the restoring force (see section 1.2) so it is reasonable that wood, which bends more easily than steel, will have a weaker restoring force, and bounce back more slowly than steel. In the case of strings, the restoring force is set by the tension applied on the string, i.e. the higher the tension, the higher the frequency of the fundamental. The difference is that in musical instruments, the rods are not under any tension. Here we will discuss a case of interest to percussion instruments, namely a rod free at both ends.

Figure 6.6(a) shows the fundamental mode for a rod that is free at both ends. This can be realized, for example, by suspending the rod from a string, which is the basic principle of chimes, including common wind chimes. We note that the fundamental mode has two nodes, the second mode will have three nodes, and so on. The frequency of the second mode is not an integer multiple of the frequency of the fundamental. Instead, it is 2.78 times the frequency of the fundamental. The amplitude of the second mode may be high, even higher than the amplitude of the fundamental, if the rod is struck midway between the center and one of the free ends. The tuning fork, shown in figure 6.6(b) is a variation of the rod free at both ends. Essentially, a tuning fork is a bent rod therefore the fundamental mode has two nodes, as in figure 6.6(a), located near the bottom of each prong. Tuning forks were widely used in the past for tuning musical instruments.

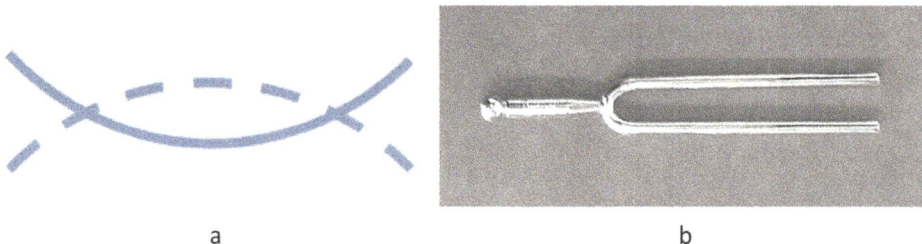

a   b

Figure 6.6. (a) Fundamental mode of a rod free at both ends. (b) Tuning fork.

6.11 Harmonics, partials, and overtones

An **overtone** is any mode above the fundamental. For instance, the second mode in a string is the first overtone. **Partial** is any mode, in other words, the fundamental or an overtone. So the fundamental is the first partial, the second mode is the second partial, and so on. If the overtones have frequencies that are integer multiples of the frequency of the fundamental, then we speak of **harmonic overtones,** or simply **harmonics**. Such was the case with the mode frequencies of the vibrating string, and the air column in pipes. If the overtone frequencies are not an integer multiple of the frequency of the fundamental mode, then we have **non-harmonic** or **anharmonic overtones**. Such was the case of the rods discussed in the previous section. Anharmonic overtones occur also in plates and stretched membranes, as will be discussed in section 11.5.

6.12 Resonance and damping

An oscillator is a system that can oscillate; a swing, a string, etc. If a swing is given one push, or a string is plucked, then we have a free oscillator. A free oscillator has its characteristic or **natural frequency** (or frequencies). The natural frequency of a swing is determined by the length of the swing. Similarly, a string has a number of natural frequencies, its vibrational modes, each with its own frequency as discussed in section 6.3. For simplicity we will use the swing as our example in what follows. When the swing is pushed again and again, the swing becomes a **driven oscillator**. Children discover early on that they can drive the oscillator themselves by pumping: leaning back and extending the knees when the swing is moving forward, then bending the knees and leaning forward when the swing is moving back. Either way, pushing or pumping does not necessarily make the amplitude of the oscillation larger. This can be achieved only if the swing is driven with proper timing, which is different for each swing.

As it turns out, if the driving frequency is low compared to the natural frequency of the swing, the amplitude of the oscillation will be small. The same happens if the frequency of the driving force is high compared to the natural frequency of the swing. If the frequency of the driving force is close to the natural frequency of the swing, then the amplitude of the oscillation can become very large. In this case we have **resonance**. Exactly how large the amplitude can get depends on the energy loss associated with the motion of the swing. For instance, if the swing is too low and the child's feet are dragging on the ground on every cycle, the amplitude at resonance will be smaller. This energy loss could arise from friction, air resistance, or other factors, which we will refer to as **damping**. How close the driving frequency has to be to get the maximum amplitude depends on the amount of damping present.

Figure 6.7 shows the amplitude of the oscillation for three different cases. The blue line shows the amplitude of the oscillator with very little damping. The red line shows the amplitude when damping is increased ten times, and the green line shows the amplitude when damping is increased 100 times. The horizontal axis is the ratio of frequency of the driving force divided by the natural frequency of the oscillator. Therefore, at the value '1' the oscillator is driven at its natural frequency. The driving frequency is lower than the natural frequency for values less than one and

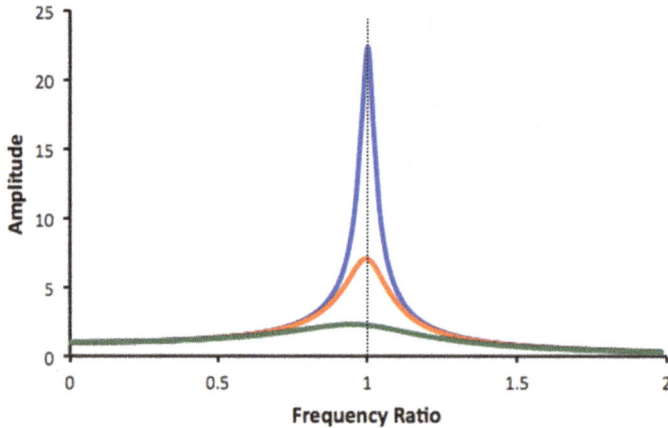

Figure 6.7. Amplitude of a driven oscillator for three values of damping. The peak is narrow and high when damping is low.

vice versa. The vertical axis shows the amplitude of the oscillation in arbitrary units. For the case of low damping (blue line) the peak value corresponds to an over 20-fold increase, compared to the values at the left end of the graph. The peak occurs when the driving frequency is equal to the natural frequency. At this point we have resonance and the amplitude of the oscillation assumes a maximum value. For the intermediate case (red line) the peak is lower and the curve is broader, i.e. the resonance is not as sharp or strong. For high damping (green line) the peak is very broad and very low. Note also that the resonance occurs at a driving frequency slightly lower than the natural frequency of the oscillator. Because of the associated energy loss, one may be tempted to think that damping is an undesirable quantity. This is not always the case. For instance, the diaphragm of a loudspeaker (see section 8.7) is an oscillator driven by the output of the amplifier. Without damping, the diaphragm would respond strongly to a certain frequency and much less to the other sound frequencies. The high amplitude of oscillation at resonance may be enough to destroy the loudspeaker. The presence of damping makes the frequency response wider and diminishes the amplitude of the resonant frequency.

6.13 Examples of resonance

Resonance plays a significant role in musical instruments and indoor acoustics. The vibrational energy of a string does not transfer efficiently to the surrounding air. Therefore, the vibrational energy is first transferred to a box (e.g. the body of a violin) and then to the surrounding air. If the box happens to have any sharp resonances, the corresponding frequencies will sound louder, which is undesirable for an instrument. Undesirable resonances may also occur in rooms and auditoriums. Closed spaces behave like pipes closed at both ends (see section 6.9) in that they can sustain a set of vibrational modes. As a result, some frequencies may sound louder than intended. This problem is particularly noticeable with car sound systems, where the low frequencies can excite vibration in parts of the car interior.

Similarly, in small rooms resonances can build up at the vibrational modes of the room, and set windowpanes into vibration.

One can use a guitar, for example, to actually see resonance in strings. The effect is more obvious if one uses the two top strings. If the guitar is tuned, pressing one finger at the fifth fret of the top (E) string corresponds to the same tone as the open fifth string (A). When the top string is plucked, the A string will begin to vibrate as well. The vibrations of the top string are transmitted through the body of the guitar to the A string, and since the frequencies match, we have resonance. If the finger is at the sixth fret, plucking the top string has no effect on the A string, because the frequencies are not the same, therefore, there is no resonance.

A related example occurs in some string instruments. The viola d'amore and the Indian sitar have a set of 'sympathetic' strings. The musician does not play these strings. Instead, they are tuned to resonate with some of the tones that are produced by the played strings, and thus sustain the played tone a bit longer.

The amplitude of the oscillation at resonance can get high enough as to damage the oscillator; hence, accounts of breaking wine glasses with one's voice. The vibrational modes of a wine glass are within the range of the human voice, and power levels over 100 dB (see section 4.4) at the resonance frequency of the glass can shatter the glass, according to accounts[3].

6.14. Further discussion

Thick strings

The mode frequencies described in section 6.2 apply to perfectly flexible strings. In practice strings have some stiffness, especially the thicker ones. Stiffness acts as a restoring force that adds to the tension applied on the string. Looking at figures 6.1–6.3 we note that the fundamental has a more gentle bend. For the overtones the string has to do more bending; in other words, the effect of stiffness is more significant for the harmonics. As a result, the restoring force for the overtones is higher. A larger restoring force means that the string oscillates faster. So, if the fundamental is 110 Hz, the first overtone may be 220.5 Hz rather than 220 Hz, and the second overtone may be 331 Hz rather than 330 Hz. This means that the overtones do not follow the harmonic series. Thick strings are necessary for producing low frequency tones in musical instruments, like pianos, guitars, etc. To avoid the effects of stiffness, the strings use wound wire around a thin metallic or nylon core. The wire winding adds thickness to the string and at the same time keeps it flexible.

Inhaling helium

Many of us had the opportunity to observe a science demo where one produces a 'cartoon' voice by inhaling helium gas. The human voice is essentially produced in the vocal tract, which consists of the larynx, the pharynx, and the oral cavity. The vocal tract acts like a semi-closed pipe that sustains standing waves. The dimensions

---

[3] The author has not witnessed such an event.

of the pipe determine the wavelengths of the modes. Now, the product of the frequency times wavelength must equal the speed of sound. If the vocal tract is filled with helium, the wavelength of the sound remains the same as in air. But the speed of sound in helium is about 1000 m s^{-1}, which is about three times larger than the speed of sound in air. Therefore, the frequencies produced when the vocal tract is filled with helium gas are three times larger than the frequencies produced in air. Hence the cartoon voice.

6.15 Equations

Speed of wave in a clamped string

The speed c of a wave in a string is given in terms of the tension of the string F and its linear density d (i.e. the mass of the string divided by the length if the string) by

$$c = \sqrt{F/d}.$$

Mode frequencies in a clamped string

$$f = \frac{n}{2L}\sqrt{F/d}$$

where L is the length of the string and $n = 1, 2, 3$, etc.
From the above equation it follows that:
- The fundamental mode ($n = 1$) has the lowest frequency.
- Longer strings produce lower frequencies.
- The frequency increases with tension.
- The frequency decreases if the linear density is higher, which means that thicker strings (of the same material) produce lower frequencies.

6.16 Questions

1. (a) What is the longest wavelength a standing wave can have in a 0.6 m long open pipe.
   (b) What is the longest wavelength a standing wave can have in a 0.6 m long string clamped at both ends?
   (c) Is the frequency of the sound produced by the pipe and the string the same? Explain your answer.
2. (a) An open pipe has fundamental frequency of 50 Hz. List the frequencies of the first 3 harmonic overtones.
   (b) A semi-closed pipe has fundamental frequency of 50 Hz. List the frequencies of the first 3 harmonic overtones.
   (c) Which of the two pipes is longer?
3. A certain instrument produces the following set of frequencies: 100, 200, 300, 400 Hz and so on.
   (a) Which frequency is the fundamental?
   (b) Which frequencies are the overtones?

    (c) Are the overtones harmonic? Why, or why not?

    (d) Which frequency is the second harmonic?

    (e) Which frequency is the second overtone?

4. A certain instrument produces the following set of frequencies: 100, 202, 303, 404,...Hz.

    (a) Are the overtones harmonic? Why or why not?

    (b) Is this a pure tone? Explain.

    (c) Which frequencies are the partials of this tone?

Chapter 7

Analog and digital signals

7.1 Introduction

In everyday life we use the **decimal system** to express numbers. The decimal system uses the **digits** 0–9 in each place, and is based on the powers of 10, i.e. 1, 10, 100, 1000, and so on[1]. For example, in the decimal system the four-digit number 2193 means (reading from right to left) 3×1, plus 9×10, plus 1×100, plus 2×1000. This is not the only possible number system. As will be explained below, the so-called **binary system** is more suitable for computers. The binary system is based on powers of 2, i.e. 1, 2, 4, 8, and so on. For example, in binary the number 1101 means (reading from right to left) 1×1, plus 0×2, 1×4, plus 1×8, which equals $1 + 0 + 4 + 8 = 13$ in decimal notation.

The binary digit or **bit** for short is the basic unit in the representation of binary numbers, and the system uses the digits 0 or 1 in each place. Using one bit we can represent the number 0 or the number 1. Using two bits we can represent four numbers

---

[1] Note that any number raised to the power 0 equals 1. For example, $10^0 = 1$ and $2^0 = 1$, and so on.

(including 0), namely the number 0 (as 00), the number 1 (as 01), the number 2 (as 10), and the number 3 (as 11). Obviously, many more bits are needed to represent numbers larger than 3. Following are the first eight (including 0) decimal integers in binary representation using three bits, with the decimal number in parentheses:

000 (0)	001 (1)	010 (2)	011 (3)
100 (4)	101 (5)	110 (6)	111 (7)

Even more bits are required to represent numbers larger than 7. For example, to represent the numbers 0–255 we need eight bits[2]. The term **byte** is commonly used to indicate an eight-bit binary number. The use of binary numbers goes beyond the representation of integer numbers. There are several schemes to 'encode' other information, including the sign (+ or −), the decimal point, characters, etc. The details of these encoding schemes are beyond the scope of our discussion.

7.2 Analog and digital signals

As discussed in section 3.2 the basic procedure of measuring sound entails the conversion of the time variation of sound pressure into a time variation of an electrical voltage. The variation of the voltage with time must be 'analogous' to the variation of the sound pressure, and this voltage is an **analog signal**. Similarly, the voltage from an amplifier that sets the 'cone' of a loud speaker into a vibration *proportional* to the voltage is an analog signal[3]. Figure 7.1 shows a *triangular* waveform that varies continuously with time, in other words, an analog signal.

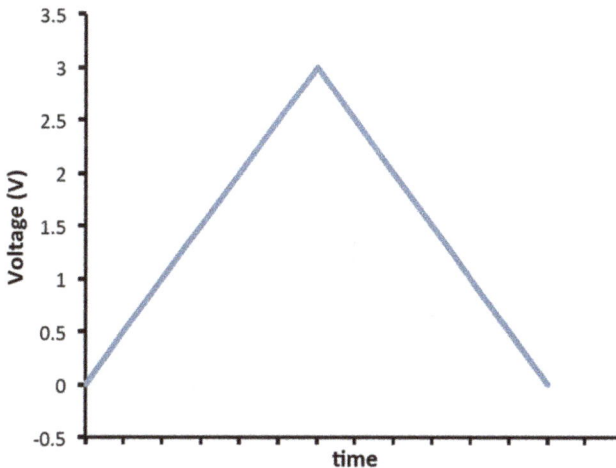

Figure 7.1. An analog signal.

---

[2] For further discussion see the section at the end of this chapter.
[3] The word analogous derives from the Greek word αναλογος, meaning proportional.

If we want to record and store this analog signal, we have to take a series of voltage measurements at regular time intervals. These data points are like 'snapshots' of the continuous variation of the voltage. To take a snapshot of the voltage, we need to stop the action long enough to allow the measuring device to acquire the voltage value. As a result, we end up with a set of **discrete** values representing the continuous analog signal of figure 7.1. We can 'summarize' the variation of the voltage using four discrete values only, namely 0, 1, 2, and 3. In this case we assign the value 0 V to any voltage lower than 0.5 V; we assign the value 1 V to any voltage between 0.5 and 1.5 V; we assign the value 2 V to any voltage between 1.5 and 2.5 V; and assign the value 3 V to any voltage above 2.5 V. In this way the triangular waveform is represented by four discrete values, as shown by the red graph in figure 7.2.

This discrete representation (red line) in figure 7.2 deviates from the original triangular form (blue line) by 0.5 V at most, i.e. we can have an error of ±0.5 V. What is interesting here is that we can describe the four discrete states using a two-bit binary number. The signal may appear very simple, but as discussed in section 3.2, for a real sound signal we need to take as many data points as required to ensure that we are not missing any 'wiggles' or other features that may be present, and this is something that we do not know beforehand.

An eight-bit binary number allows 256 states. In other words, an eight-bit binary number would allow us to use 256 discrete values, and that would diminish the error substantially, as shown in figure 7.3. The red plot is actually a number of discrete points. The inset shows a magnified view of the waveform around 1 V. Note that the deviation from the triangular waveform (blue line) is roughly ±0.006 V.

In the discrete form (figure 7.2) the signal follows the sequence 0; 1; 2; 3; 2; 1; 0. Using three-bit binary numbers (see the list of decimal to binary equivalents in the previous section) this pattern of voltages can be represented in binary by the following sequence:

000 (0); 001 (1); 010 (2); 011 (3); 010 (2); 001 (1); 000 (0).

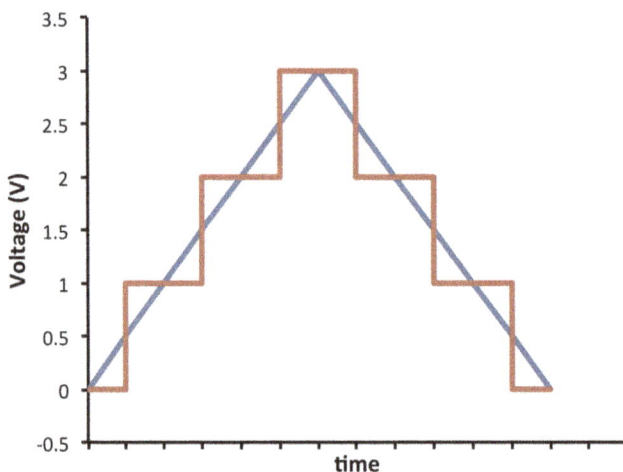

Figure 7.2. Discrete representation of a triangular waveform using four discrete levels.

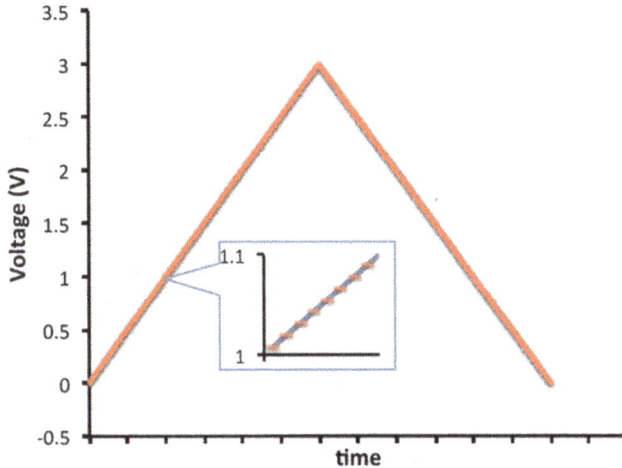

Figure 7.3. Discrete representation of a triangular waveform using 256 discrete levels. Note that the difference between the discrete and the analog values is much smaller than in figure 7.2.

This set of values gives us the analog signal in binary form. The advantage of using binary numbers rather than decimal numbers is that computers work with binary numbers. Of course, we need to have an agreement or protocol to understand this code. First of all, we need to know how many bits represent a number (in the above example we used three-bit numbers) and also the rate at which these numbers are coming. We also need to agree on how to read each binary number. For example, we can read from left to right. In terms of audio applications, most of these specifics are handled by sound cards, which are usually part of the computer system.

7.3 Analog to digital and digital to analog conversion

The heart of a computer is the so-called central processing unit (CPU). The transistor is the basic element of the CPU and of most of the other components of the computer system. Depending on how it is configured, the transistor can act as an amplifier, e.g. in sound equipment and portable 'transistor' radios. The transistor can also act as an electronic switch, and this is mainly how the transistor is used in computers. As a switch, the transistor can have two states: ON or OFF, and it can be used to switch a voltage, i.e. switch on and off a voltage of 5 V. In this case the voltage can have two levels: 0 or 5 V. The two states of the transistor (ON or OFF) can also be indicated using the values 0 and 1, which is exactly what can be represented by a binary digit (or bit).

Using a large number of transistors integrated in a very small area (about 7 billion in an area of about one square inch), the CPU can perform a huge number of logical and arithmetic operations in binary. Therefore, if we have a sound signal from a microphone (which is an analog signal) that needs to be processed by the computer, the signal must first be converted to binary or digital as it is commonly called. Once the computer has processed the digital signal, this digital signal must be converted back to analog before it is fed to an analog device, such as a loudspeaker.

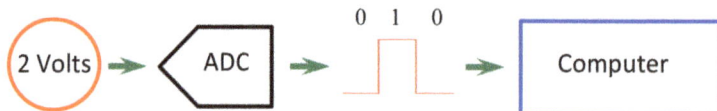

Figure 7.4. Path of a signal from an analog device (red circle) to a computer.

Figure 7.5. Path of a digital signal from a computer to an analog device.

There are two types of devices that are designed to do the necessary conversions: the analog to digital converter (ADC) and the digital to analog converter (DAC). The ADC converts all the incoming analog signals into binary code for further processing or storage by the computer. The DAC coverts the binary code and outputs the information as an analog signal.

Figure 7.4 is an oversimplified diagram showing the path of an incoming analog signal. For example, suppose that we have a three-bit ADC and that the signal level at this point in time is 2 V, which in three-bit binary is represented by 010 (see list in section 7.1). The voltage is applied to the input side of the ADC, which converts it to its digital form, i.e. 010. This number will appear at the output side of the ADC, in this case as a train of voltage pulses. This train of pulses is then routed to the CPU of the computer. During this time, the analog signal at the input may have changed, say to 3 V. Once the processing of the 2 V signal is complete, the ADC digitizes the next value, i.e. the 3 V, and so on.

Figure 7.5 shows the path of an outgoing digital signal from a computer. For example, suppose that the CPU completed an operation, and the result is 101. The decimal value of this binary number is 5 (see list in section 7.1). The digital signal is applied at the input side of the DAC in the form of voltage pulses. The DAC converts the digital signal to analog, and the result (5 V) appears at the output side of the DAC. This signal is now ready for use by an analog device, like loudspeaker, and the DAC is ready to process the next digital number from the computer.

In many cases the ADC and DAC are parts of the same unit. The **sound card** on computers has several input and output ports. The opening figure of this chapter shows a computer sound card. Typically, a sound card will have an input for low voltage signals (i.e. the signal from a microphone, which is in the order of a few thousandths of a V); this input is pink. The light blue port is for input of the so-called line level signals, i.e. from a radio or a tape player, which is in the order of 1 V. The lime green output is intended for headphones or as input to stereo amplifiers. The silver, orange, and black outputs are intended for surround-sound systems. All these input and output signals are analog signals.

7.4 Further discussion

Minimum sampling rate

In measuring or in digitizing a signal, it is important to know the rate at which we should be taking measurements, the so-called **sampling rate**. As discussed in section 3.2, the sampling rate depends on the frequency components present in the analog signal. For example, in the audible range, components with frequencies up to 20 000 Hz may be present. The minimum sampling rate is given by the **Nyquist theorem**, which tells us that the sampling rate should be at least equal to twice the highest frequency present. In other words, if the highest frequency is 20 000 Hz, then we should be sampling the signal 40 000 times per second. Most sound cards sample at 44 000 Hz.

7.5 Equations

Reading binary numbers

In the familiar decimal system, the base is 10, and the digits (0 through 9) are understood to multiply the powers of 10. Thus the five digit number 20 358 means

$$2 \times 10^4 + 0 \times 10^3 + 3 \times 10^2 + 5 \times 10^1 + 8 \times 10^0$$
$$= 2 \times 10\,000 + 3 \times 100 + 5 \times 10 + 8 \times 1 = 20\,358.$$

In binary the base is 2, and the available digits are 0 and 1. The binary number 101101 means

$$1 \times 2^5 + 0 \times 2^4 + 1 \times 2^3 + 1 \times 2^2 + 0 \times 2^1 + 1 \times 2^0 = 1 \times 32 + 0 \times 16$$
$$+ 1 \times 8 + 1 \times 4 + 0 \times 2 + 1 \times 1 = 32 + 0 + 8 + 4 + 0 + 1 = 45.$$

With six digits in decimal we can represent 10^6 ($=1\,000\,000$) numbers, namely from 0 to 999 999. In general, using N digits we can represent 10^N numbers (from 0 to 10^N-1).

With six-bit binary numbers we can represent 2^6 ($=64$) numbers, namely from 0 to 63. Using N bits in binary we can represent 2^N numbers (from 0 to 2^N-1).

It is common to use eight-bit binary numbers, meaning that we can represent 2^8 ($=256$) numbers, from 0 to 255.

In computer science, it is also common to use **hexadecimal** (**hex** for short) numbers[4]. The base in hex is 16. Therefore, we need 16 digits. This is accomplished by using alphanumeric digits. In addition to the familiar 0–9, we use A through F to represent 10 to 15, respectively. In hex the number 210B means

$$2 \times 16^3 + 1 \times 16^2 + 0 \times 16^1 + 11 \times 16^0$$
$$= 2 \times 4096 + 1 \times 256 + 0 \times 16 + 11 \times 1 = 8192 + 256 + 0 + 11 = 8459.$$

Quantization

The process of assigning discrete values (see figure 7.2 above) to a continuous (analog) signal is called **quantization**. As discussed above, the discrete form deviated

---

[4] The reader may be familiar with computer color codes, which are usually in hex.

from the analog form, and this deviation, or **quantization error** depends on the number of discrete levels introduced, which in the binary form is determined by the number of bits used. If our signal varies from 0 to A (= 3 for example in figure 7.1) and we use N bits in our binary representation, then we have 2^N *levels*, or $2^N - 1$ *steps* in between 0 and A. The difference between the digital signal and the analog signal can be half of a step at most. Therefore the quantization error can be at most $\pm 1/2 \ A/(2^N - 1)$. In the example of figure 7.3, where we use eight bits (recall $2^8 = 256$), and $A = 3$, the quantization error is $\pm (1/2) \ 3/(256 - 1) = \pm 0.006$ V.

7.6 Questions

1.   (a) Express the decimal number 6 in binary.
        (b) Express the binary number 0101 in decimal.
2. Is the voltage of the wall outlets analog or digital? Explain.
3. Refer to the opening figure of this chapter. Which of the input/output ports are digital?
4. Why do digital computers work with binary numbers?

Chapter 8

Sound equipment and components

8.1 Introduction

Sound systems have evolved dramatically in the second half of the 20th Century. The advances in electronic technologies and materials are probably more than many people could have imagined in the 1950s. The main purpose of sound systems still remains the same: to output sound of good quality, at a power level suitable for the listening environment. The earlier sound systems used vacuum tubes, which are

doi:10.1088/978-1-6817-4680-7ch8

bulky, expensive to fabricate, produce a lot of heat, and have limited lifetime, like ordinary incandescent light bulbs. Today vacuum tubes are still in use for specialized applications, but have largely been replaced by the so-called semiconductor or solid-state devices, which is what we have in computers, cellphones, digital cameras, and of course sound systems.

One of the major advantages of solid-state technology is the phenomenal reduction in size. A portable radio nowadays is smaller in volume than a single vacuum tube. A key component in sound amplification is the **transistor**, which is a solid-state device that can use a small signal to control a large electric current provided by a large power supply. In other words, a transistor creates a larger replica or amplifies the small signal. In this chapter, we discuss the basic functions of sound equipment and components. To better understand the performance characteristics of these systems, we will introduce some concepts of electricity that will be useful in the discussion that follows.

8.2 Concepts of electricity

To introduce some of the basic concepts, we use the simple case of a voltage source and a load, as shown in figure 8.1. The voltage source, which could be a battery, an amplifier, a microphone etc, drives an electric current through the load. Depending on the source, the load can be a light bulb, a loudspeaker, etc.

By convention, the direction of the electric current is from the positive terminal of the voltage source, through the load, and back to the negative terminal of the source, as indicated in figure 8.1. If the voltage source provides a constant voltage, for example a battery, then we speak of a direct current (DC) source. In sound applications the voltage can be variable, and change polarity, i.e. the positive terminal becomes negative, then back to positive, and so on. In this case, the direction of the current changes as well, and we speak of a variable or alternating current (AC). The wall outlets in houses are AC sources.

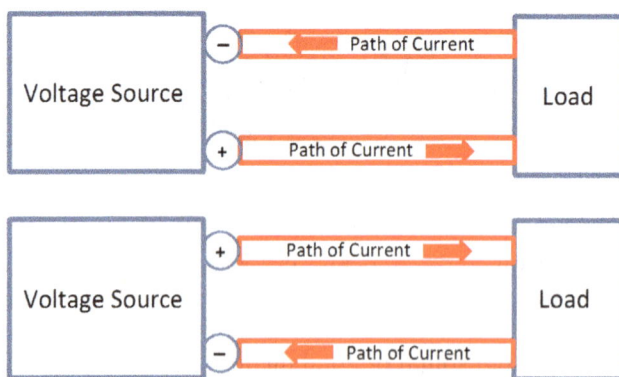

Figure 8.1. A load connected to a voltage source.

In North America the voltage varies from about $+170$ to -170 V 60 times per second, i.e. we have an AC of 60 Hz[1]. The current is usually carried by electrons, and the **electric current** is a measure of the rate of electron flow. The **voltage** is what provides the energy required for the electrons to move. As the electrons flow through the load, they slow down by collisions with the atoms in the wires or other components they may encounter along their path. The resistance of the device is a measure of the hindrance encountered by the electrons as they move through a device. During these collisions the electrons lose some of their energy, which is converted to heat.

Resistors are electronic devices that have more or less constant resistance for both AC and DC voltages. A simple relation known as **Ohm's law** relates the voltage *across* a resistor to the current *through* the resistor:

$$\text{voltage} = \text{resistance} \times \text{electric current}.$$

Ohm's law tells us that for a given supply voltage, e.g. 9 V, the product resistance $\times$ current must equal 9. For instance, we can have 1×9 or 2×4.5, or 3×3, etc. Note here that as the resistance increases the current decreases and vice versa. Running more current means more loss in collisions, i.e. more heat. Besides being a loss, heat in electrical devices can affect performance or even damage the unit. The current drawn/delivered to any unit must never exceed the values specified by the manufacturer.

The **power** delivered to a load is equal to the product of the voltage across the load, times the electric current through the load:

$$\text{power} = \text{voltage} \times \text{current}.$$

If the load is a resistor this power is 'dissipated', meaning it is converted to heat. A different load, such as a loudspeaker, would convert some of the power delivered into sound.

Besides collisions, which occur for both AC and DC signals, for AC currents there are two additional mechanisms that affect the flow of electrons through a device. In the first mechanism the electrons are temporarily stored. A device that stores electric charge is called a **capacitor**, which typically consists of two metallic films separated by an insulating layer. The amount of charge that can be stored by a capacitor depends on the separation or **gap** between the two films.

A second mechanism that affects the flow is the interaction of electrons with a magnetic field. This happens for example when the electrical current passes near a magnet. Alternatively, the electric current can interact with the magnetic field that is created by the current itself[2]. In either case, this interaction tends to counteract any change in the value of the current. This phenomenon is known as **induction**. Commonly an **inductor** is a coil of thin insulated copper wire. The coil may or may not be wrapped around a magnetic material. Both inductors and capacitors

---

[1] The amplitude is 170 V. Commonly the voltage is referred to as 120 V, AC. The 120 is the so-called root mean square (rms) value. The rms value is the amplitude (170) divided by the square root of 2.
[2] Any wire carrying an electric current creates a magnetic field.

affect the flow of current, but contrary to resistors, they do not waste much energy by generating heat. It is common to have an AC device (or load) where all three mechanisms are present at once. In such a case we speak of the **impedance** of the load, which is analogous to the resistance in DC. Unlike the resistance, the impedance depends on the frequency of the signal.

A voltage source also has some *internal* impedance. As a general rule, the impedance of the source must be lower than the impedance of the load. Otherwise the voltage across the load will be smaller than the voltage that the source can deliver. This fact is significant when connecting a microphone (source) to an amplifier (load) or a loudspeaker (load) to an amplifier (source).

Finally, a note about units:
- Voltage is measured in **Volts** (V) or mV (1/1000 of 1 V).
- Electrical current is measured in **Amperes** (A) or mA (1/1000 of 1 A).
- Resistance and Impedance is measured in **Ohms** (Ω) or kΩ (1000 Ω).
- Power is measured in **Watts** (W) or mW (1/1000 of 1 W) or kW (1000 W).

8.3 Filters

Filters are devices that can block a range of frequencies. The voltage signal to be processed is applied on the input side, and the filtered signal appears on the output side of the filter. Depending on the range of frequencies that are blocked, we have four types of filters: low-pass, high-pass, band-pass, and notch. An ideal **low-pass** filter will completely block all frequencies higher than a **cutoff** frequency. Similarly, an ideal **high-pass** filter will completely block all frequencies lower than a cutoff frequency. The **band-pass** filter will block all frequencies that are not within a certain range. Finally, a **notch** filter will block only the frequencies that are within a certain range. Figure 8.2 shows the response curves for 'ideal' filters. The vertical axis

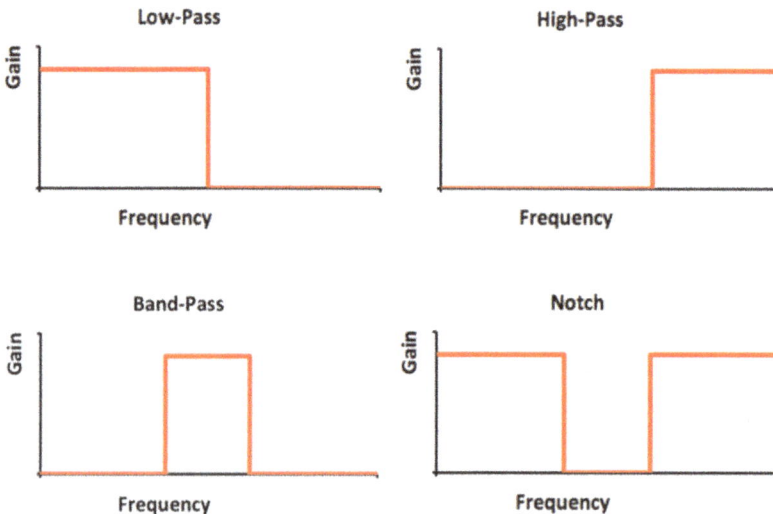

Figure 8.2. Response curves for ideal filters.

usually indicates **gain**, which is defined as the ratio of the output voltage to the input voltage, and is usually measured in dB (see sections 4.4 and 4.14). In the figures, low gain means that the signal is blocked and vice versa. In practice, the response curves of real filters do not have corners, and the transitions from the passed to the blocked frequencies are smooth[3].

The cutoff frequencies and the location and bandwidth of the pass-band and notch filters can be selected by the designer, and can be 'tunable' as well, so one can have a knob that selects the cutoff frequencies, as is done with radios. In connection with loudspeakers, a very useful device is the **crossover** network of filters, which divides the signal going to the loudspeaker system into two or more ranges (for example, low, mid-range, and high) and directs them to the proper speaker: low frequencies to the woofers, high frequencies to the tweeters, etc, as discussed in section 8.7.

8.4 Microphones

Microphones are devices that convert the pressure variation of sound waves into variations of voltage. A device that converts one physical quantity (such as pressure) to another (such as voltage) is called a **transducer**. The element or **capsule** of a microphone most commonly consists of a flexible **diaphragm** that is set to motion by the sound wave. The next step is to convert this motion into electric voltage. One way is to attach a small wire coil to the diaphragm, and have the coil move in an out of a magnetic field. Recall from the discussion of the inductor (section 8.2) that magnetic fields interact with the electrons in a wire, and this interaction *induces* a voltage between the two ends of the coil. If the coil is moving, a voltage replicating the motion of the diaphragm will develop between the two ends of the coil. This is the microphone signal. This type of microphone is called a **dynamic** microphone.

Another design uses a piezoelectric material. When pressure is applied between two sides of a piezoelectric material, a voltage develops between the two sides. In **piezoelectric** or **crystal** microphones, the diaphragm is attached to one side of a piezoelectric crystal, and the pressure applied by the diaphragm on the crystal creates a voltage that mimics the pressure from the sound wave. In another design, a metallic diaphragm is arranged parallel to a rigid metallic plate, the 'back-plate'. The pair of metal plates essentially forms a capacitor. A positive DC voltage (e.g. from a battery) is applied to the diaphragm. If the diaphragm moves in response to an incoming sound wave, the gap between the plates of the capacitor (and its ability to store charge) will change accordingly. In other words, the charge stored by the capacitor will change, thus causing electrical charge to move (i.e. causing an electric current to flow) to or from the battery. The current generated replicates the motion of the diaphragm. This type of microphone is the **condenser** microphone. There are other types of microphones, such as the electret, the ribbon, etc, which are not discussed here.

---

[3] See discussion at the end of this chapter.

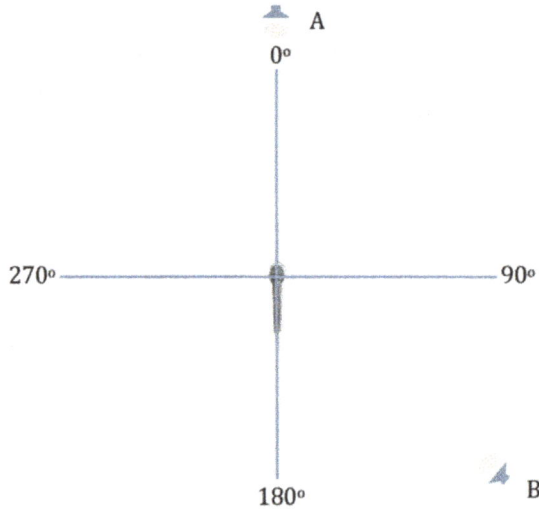

A
0°

270° ———————————— 90°

180°   B

Figure 8.3. Two loudspeakers (A and B) producing identical sounds. The line 0–180° is the axis of the microphone, which is located at the center of the figure.

One important characteristic of a microphone is the directionality or **directivity**. For example, suppose that we have a microphone and two loudspeakers, A and B. Speaker A is located on the axis of the microphone (at 0°) and B is at an angle from the axis (figure 8.3). The sounds from the two speakers are identical. One can then ask: at what distance should we place speaker B so that the signal picked up by the microphone is the same as the signal from speaker A? Depending on the answer, we can have several types of microphones.

An **omnidirectional** microphone will pick up equally well from all directions, as long as the distance of B is the same as the distance of A from the microphone. In other words, we can place speaker B anywhere on a circle centered on the microphone and get the same signal from the microphone. The so-called pickup pattern for an omnidirectional microphone is shown in figure 8.4(a). A very common pickup pattern is shown in figure 8.4(b). This is the **cardioid** pickup pattern. The microphone will produce the same signal for a speaker located anywhere on the red line. The same applies for all locations on the green curve and on the blue curve, which is the farthest from the microphone. Note that in the forward direction (at and near 0°) the pickup is much better. If we set the speaker to the sides (at 90° or 270°) we need to bring the speaker much closer to the microphone to produce the same level of signal from the microphone. The microphone will hardly pickup any signal from a speaker set behind the microphone (at 180°).

Clearly the cardioid pattern favors the forward direction. This is an example of a **unidirectional** pickup pattern. An extremely unidirectional pattern is the so-called **shotgun** pickup pattern, where the microphone picks up only sounds coming along its axis (i.e. at 0°). The pickup pattern of a microphone is basically determined by the design of the housing of the microphone capsule. Which pattern is best depends on the application. For example, on a theater stage, the omnidirectional would pick up

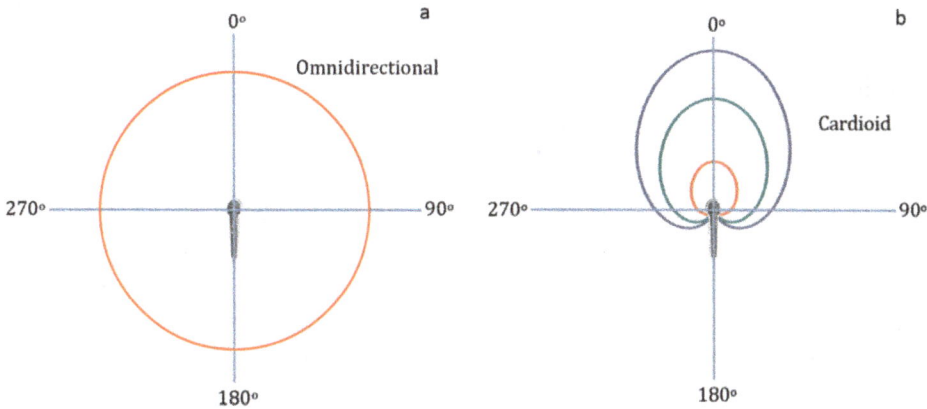

Figure 8.4. Pickup patterns for (a) omnidirectional microphones and (b) cardioid microphones.

the voices of all actors. For singing it may be best to use a cardioid, so as not to pick up sound from the nearby instruments. A shotgun pattern would work best for a reporter trying to record the voice of one person in a crowd, and so on.

Another important characteristic is the **microphone impedance**. If the impedance of the microphone is over 20 kΩ we have a high impedance microphone, such as a condenser or piezoelectric microphone. Low impedance microphones, such as a dynamic microphone, have impedance less than 1 kΩ. When connecting a microphone to an amplifier, the microphone is the voltage source and the amplifier is the load. As discussed in section 8.2, we need the impedance of the load to be higher than the impedance of the source. The input impedance of an amplifier is typically around 1–2 kΩ. So, when connected to a microphone cable the signal from high impedance microphones drops, and the drop affects mostly the high frequencies in the signal. So, unless other precautions are taken, a high impedance microphone should not be connected to a cable longer than 6 m (about 20 ft). The low impedance microphones do not have this problem. Generally, the voltage output from a microphone is very weak, in the order of a few millivolts, and requires amplification. The first stage is the so-called pre-amplifier, which brings the microphone voltage signal to the proper level.

8.5 The amplifier

The loudspeakers in a home sound system are typical in the range of 25–100 W, while the signals coming from a CD player, a tape player, etc, are usually very low. For example, the voltage output of a CD player is about 1 V and the current in the order of mA[4], meaning that the power output is about 1 V × 1 mA = 1 mW. This signal level needs be amplified by about a million times in order to drive a 100 W speaker. This amplification can be accomplished by increasing the current or the voltage or both. For safety reasons, high voltages are not desirable, especially in home systems. Typically, the voltage to a loudspeaker is in the order of 10 V,

---

[4] See unit abbreviations in section 6.2.

although some concert speakers operate with voltages in the order of 100 V. Therefore, the power boost must come mainly from increasing the current by at least 1000 times. The so-called **power amplifier** (or **power amp** for short) does this final boost. Processing currents over a few mA generates heat, i.e. loss of power, therefore any adjustments of the signal to be amplified must be done before the signal reaches the power amp; for example, adjusting the volume, the bass, or the treble. These controls occur in the pre-amplification stage, and can be relatively simple for most home systems, or very complex, as in studio sound recording and professional sound systems.

In most home sound systems a **pre-amplifier** (**pre-amp** for short) is part of the **receiver**. The receiver in home stereos includes the **tuner** (for AM/FM radio) and a power amplifier, which feeds the loudspeakers[5]. In some home systems, and most of stage and recording sound systems, the pre-amp is a separate unit. First the pre-amp brings the signal levels from all inputs to the right level for the power amplifier to handle. Recall that the signal from a microphone is in the order of millivolts, while signals from CD players, tape players, etc, are in the order of 1 V[6]. All these signals together go into the power amp, so they must be at the same level. Adjusting the volume is done in the pre-amp stage, and can be as simple as adjusting the voltage level using a variable resistor.

In terms of frequency control, a simple pre-amp has knobs for tone control, i.e. adjusting the mix of low and high frequencies, the bass and the treble, respectively. This is achieved by taking the signal through high-pass and low-pass filters before feeding the signal to the power amp (see section 8.3). An **equalizer** uses a combination of filters to achieve finer control of the frequency bands. In this case, one can have three (or many more than three) frequency bands that can be adjusted separately.

Another function of the pre-amp is to add or mix signals. For example, adding the signals coming to the pre-amp from two or more different sources, and adjusting the proportions in the mix. For example, a DJ may have a microphone and two or more music sources (CD player, record player, smartphone, etc). The DJ uses an **audio mixer**, which is a pre-amp that can mix the sound from all the inputs, and by adjusting the level of each, can make smooth transitions from one song to the next. Similarly, a pre-amp mixes the voice and the music in a karaoke machine. More elaborate units used in performances and recording studios can mix inputs from up to about a hundred different inputs.

8.6 Amplifier characteristics

A basic requirement for a good audio amplifier is to output a voltage signal that closely replicates the variation of the input signal. So, if the amplitude of the input signal doubles in 1 s, then the output signal should do the same, otherwise the sound comes out distorted. To avoid **distortion** the amplifier must be **linear**. Typically, the

---

[5] See discussion of AM/FM bands in section 1.6.
[6] See also the discussion in section 7.3.

components of amplifiers, mainly the transistors, have a certain input range for which the response is linear. If the input exceeds this range, the output will be distorted.

A response curve is a plot of the magnitude of the output signal versus the input signal; meaning that if we know the input value (horizontal axis) we can read the output from the vertical axis. Figure 8.5 shows an 'ideal' response (green line) and a more realistic response curve (red line). We note that for small input signals (less than about 0.04 V in this example) the red curve is fairly linear. For an input of 0.02 V, the output is 2 V whether we use the red or green line. This is not the case for higher input signals. If the input is 0.1 V, the red curve reads about 7.5 V and the green line reads 10 V.

A non-linear response adds harmonics (which are not present in the input signal) to the fundamental frequency of the tone (see section 5.11). The fraction of the power of the added signal (from the harmonics) to the signal of the fundamental is the **total harmonic distortion (THD)**. The THD is measured in %, and a typical manufacturer specification may read: less than 1% THD. Of course, the higher the percentage, the lower the quality of the amplifier. A quality unit should have THD less than 0.1%.

The ratio of the output to the input voltage is the **voltage gain**. For example, in the linear range of figure 8.5, a 0.02 V input gives an output of 2 V. Therefore, the gain in this case is 2/0.02 = 100. Similarly, we can define the **power gain** as the ratio of the power output to the power input. It is common to express the gain in decibels (dB) as discussed in section 4.4. A power gain of 20 dB means that the amplifier boosts the power of the input signal by a factor of 100, and a gain of 60 dB means a power boost by a factor of 1 million. Of course, a high gain is desirable, but most

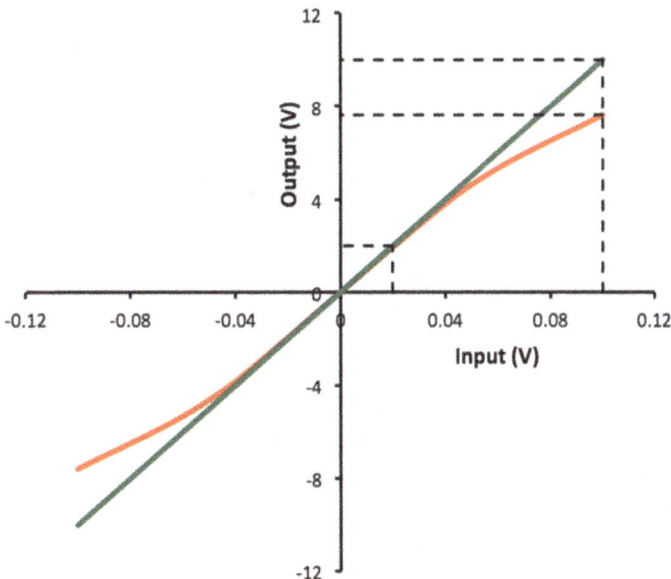

Figure 8.5. Linear (green line) and partially linear (red line) response of an amplifier.

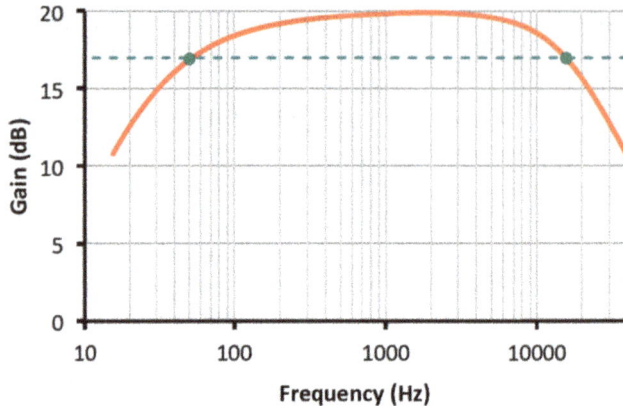

Figure 8.6. Frequency response curve for an amplifier. The green dots at 50 and 15 000 Hz indicate the frequencies where the gain drops by 3 dB (50%).

importantly the gain should be the same for a wide range of frequencies, and preferably covering the entire audible range, from about 20 Hz to 20 kHz.

Figure 8.6 shows a frequency response curve. Note that the frequency axis is logarithmic (i.e. the steps represent factors of 10). The first vertical grid line after 10 Hz corresponds to 20 Hz, the next to 30 Hz, and so on. The response curve levels out at 20 dB, and rolls-off at the two ends of the frequency range. The green dotted line marks the 17 dB gain level, and the two green dots indicate the points where the gain drops to 17 dB, i.e. drops by 3 dB from the highest value. In terms of power, a 3 dB drop means that the power drops by 50%, and this happens at 50 Hz in the low frequency side, and 15 000 Hz in the high frequency end of the diagram. These two frequencies define the useful range of the amplifier, the so-called **bandwidth** of the amplifier. Usually the specification is given as '30 Hz–16 kHz ± 3 dB', meaning that that in the range of frequencies stated the gain for some frequencies may be at most 3 dB above or below the average. A 3 dB difference means a factor of two in power. In other words, some frequencies are amplified twice as much as the average (+3 dB) or half as much as the average (−3 dB).

All electronic devices are susceptible to **noise**, usually from wires, from the supply voltage, from each other, etc. This is how we can tell that a stereo is on, even if there is no music playing. Naturally for any output signal to be useful, it should be louder than this unavoidable noise. The **signal to noise ratio (SNR)** compares the maximum output signal to the noise signal. For example, SNR = 100 dB means that the maximum output is 10 billion times higher than the noise. Recall from section 4.4 that a sound power level of 100 dB above any audible noise brings us close to the upper limit of what our ears can handle without damage.

Related to the noise level is the **dynamic range** of the amplifier, defined as the difference (in dB) between the maximum power level and the noise. For example, if the maximum output is 100 dB and the noise is 30 dB, then the dynamic range is 100 − 30 = 70 dB. A high dynamic range allows a better separation of what is supposed to sound louder in a piece of music, and what is supposed to sound softer, yet remain audible above the noise level.

An amplifier converts the power supplied by a power source to a useful audio signal. Audio amplifiers are not very energy efficient devices. The **efficiency** is measured as the output power of the amplifier to the power the amplifier draws from the power supply unit. The efficiency typically ranges from 10 to 50%. For example, an amplifier delivering 50 W at the output and has 20% efficiency actually uses 250 W total power. It outputs 50 W to the loudspeakers and 200 W as heat. The output power rating of audio amplifiers measures the power of the audio signal at the output, and for home units it can range from about 20 W to over 100 W, and much higher for professional units used in concerts. This power is delivered to the speaker system, and a very important step is to match the power rating of the amplifier to the power required to run the speakers. Generally, the power rating of the amplifier should be equal to the power rating of the speakers. If the amplifier rating is lower, then the speakers will not perform to their best. On the other hand, if the amplifier rating is much higher than the speakers' rating, the speakers will be damaged.

8.7 Loudspeakers

The loudspeaker's function is in a way the reverse of the microphone's function. The speaker converts an electrical signal from the amplifier to a sound wave. The microphone converts a sound wave into an electrical signal. In some cases, as in two-way radio units, a speaker is used both as a microphone and as a loudspeaker. The most common type for home audio systems is the **dynamic loudspeaker**, which will be described in this section. The output signal from the amplifier is connected to the voice coil, which consists of several turns of thin wire. The voice coil is suspended in a way that allows it to move back and forth in the bore of a magnet. The input signal from the amplifier produces an electric current through the coil, which in turn produces a magnetic field around the coil. The magnetic field of the coil interacts with the magnet, which causes the coil to move back and forth in a pattern that mimics the input signal. The sound coil is rigidly attached to a diaphragm that sets the surrounding air into vibration, i.e. it produces a sound wave analogous to the input signal. The diaphragm is usually cone shaped, and is attached to the body of the speaker by a flexible suspension, as shown in figure 8.7.

The range of back-and-forth travel of the sound coil defines the **excursion** of the speaker. At low frequencies, the diaphragm moves slowly, therefore the rate at which energy is delivered to the surrounding air is slow. Recall that power is the rate at which energy is delivered (see section 1.5). This means that to achieve a certain level of power at low frequencies, a larger amount of air must be moved by the diaphragm compared to higher frequencies. Low frequencies need a large diaphragm, with long excursion. On the other hand, larger diaphragms cannot move fast enough to follow the high frequencies effectively. Therefore, a speaker system usually includes three different speakers, or **drivers** as they are often called. A small driver for the high frequencies (over 6 kHz) is called the **tweeter**. The **mid-range driver** covers the mid-range frequencies (about 1–6 kHz). The **woofer** covers the low

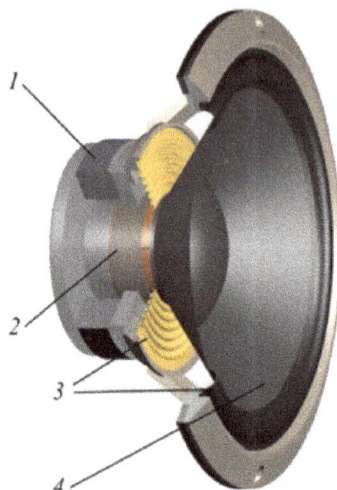

Figure 8.7. Cutaway view of a loudspeaker. 1. Magnet. 2. Sound Coil. 3. Suspension. 4. Diaphragm. Image credit: Svjo (Own work) [CC BY-SA 3.0 (http://creativecommons.org/licenses/by-sa/3.0)].

frequencies. All three drivers are usually housed in the same enclosure, thus covering the entire range of audible frequencies in most cases. A **sub-woofer** is a low frequency (below 200 Hz) driver and has its own box and own separate amplifier, and is usually placed on the floor facing down. The size of the cone is smallest for tweeters (2–5 cm diameter) and about 20–50 cm for woofers and subwoofers. As the different drivers work with different frequency ranges, a **crossover** network of filters is used to route the different frequency ranges to the appropriate driver. In some applications, covering the entire audible frequency range is not so important (for example, computers, TV, etc). These units can use a single mid-range speaker (10–20 cm cone) and usually have a double cone to enhance the higher frequencies.

Amplifiers have labels specifying the impedance of the speakers to be used, usually 4–16 Ohms. Recall that according to Ohm's law (see section 6.2) a low resistance (for DC) or impedance (for AC) load (which is the speaker in our case) draws more current from the voltage source (which is the amplifier); meaning that we should not connect a speaker with impedance lower than the minimum specified for the particular amplifier, otherwise the speaker will be driven beyond its excursion range, which will cause distortion of the sound. Worse yet, drawing high current from the amplifier causes heating of the transistors, which in excess can damage the amplifier. This can happen when two or more speakers are connected directly to the same output terminals of the amplifier.

Loud speakers are not very efficient devices. Typically, they convert to sound only about 1%–5% of the power delivered by the amplifier. In other words, if 100 W of power are delivered to the speaker, the sound wave coming from the speaker will have power of 1–5 W, and 95–99 W are converted to heat. This means that loudspeakers tend to get hot! Transferring the energy from the cone to the surrounding air is inherently inefficient, and that is the main reason for the low efficiency of speakers.

Headphones are usually miniature dynamic type speakers. They require very low power, in the order of mW, and therefore no power amplifier is needed. As they are at close proximity to the eardrum, they do not require a large diaphragm for low frequencies. So, their output frequency range can go down to 40 Hz or less. The most important concern with earphones is the level of sound reaching the ear. As they get so close to the eardrum, the entire output power is delivered into the ear, and the sound level can easily reach 80 dB, or even exceed 100 dB, which can cause ear damage, as discussed in section 4.4.

8.8 Further discussion

Amplifier power specifications

Amplifier specifications can be misleading at times. One must keep in mind that for AC signals, the power is also oscillating from a minimum (zero) to a peak value. Therefore, it is the average value of the power that is important. The peak value is actually twice the average value, therefore listing the peak value may be misleading the consumer. If the amplifier has more than one channel then it is important to know whether the power listed is total power or power per channel. In the case of two channels, the listing may be abbreviated as '100 Watts × 2', meaning 100 W per channel. Note that the power delivered depends on the load. A load of low impedance draws more power. For example, the power output to a 4 Ohm speaker may be 120 W but for an 8 Ohm speaker, the output power may be only 75 W. Therefore, the power specified must refer to the speaker impedance, for example '50 W to 4 Ohms'. The amplifier must be able to output the specified power level continuously, and the appropriate specification should clearly indicate 'driven continuously'. Also, the power must refer to the entire frequency bandwidth, not just a single frequency, say 1000 Hz. Finally, the THD should refer to the power level stated. Not listing the THD is a sign of low quality.

Real filters

Filters are widely used in many applications, including TV and radio receivers, computers, and audio systems. A very common use is to remove line noise. The 60 Hz frequency from the line voltage discussed in section 8.2 often finds its way to the output of audio systems, and results in a low frequency 'hum'. This unwanted signal could be blocked using a high pass filter. Computer input and output cables, like those shown in figure 8.8 use filters to block high frequency noise.

The characteristics shown in figure 8.2 correspond to ideal filters. Figure 8.9 shows the graphs for real filters. Note that the frequency axis is logarithmic.

Loudspeaker enclosures

Loudspeakers used in home systems usually have the three drivers (tweeter, mid-range, and woofer) housed in the same enclosure. The enclosure must be very rigid, with

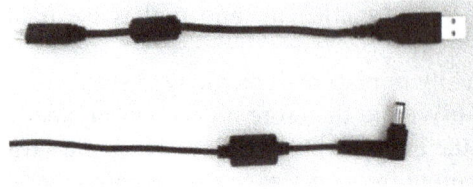

Figure 8.8. Computer cables with in-line filters. The filters are cylindrical in shape.

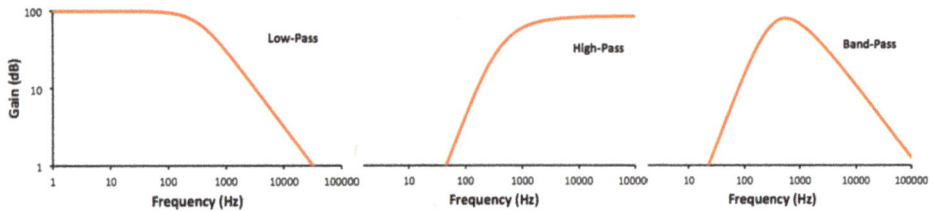

Figure 8.9. Response curves for real filters.

braces on the inner side of the enclosure the keep the sides from vibrating. The enclosure is usually filled with soft material to absorb the sound wave produced by the backside of the cone, which is 180 degrees out of phase compared to the wave from the front side (see section 2.8). In some designs, the loudspeaker enclosure is almost airtight—this is so-called air-suspension, or acoustic suspension. Another design includes an open tube in the enclosure. This tube is usually mounted at the bottom of the front side, and helps to extend the system's performance at very low frequencies. This so-called bass-reflex design can 'tune' the phase of the sound produced by the back of the diaphragm and increase the sound output from the loudspeaker system.

Connecting loudspeakers

Loudspeaker connecting points on stereo amplifiers and connecting cables are color-coded and for good reason. The signal from the amplifier to the loudspeakers is an alternating (AC) signal, i.e. the 'plus' and 'minus' alternate. If correctly connected, the diaphragm of both loudspeakers should move forward when the signal is positive, and move backwards when the signal becomes negative. If the wire connections of one speaker are reversed, then the diaphragm of one speaker will be moving forward while in the other speaker is moving backwards, i.e. the motion will be 180 degrees out of phase. If we have a certain sound wave coming from both speakers simultaneously the sound waves from the two loudspeakers will cancel each other to some extent, and the sound level will be reduced.

A high impedance load is actually a 'light' load. For example, a 16 Ohm speaker is a lighter load than a 2 Ohm speaker. When the cables of two 4 Ohm speakers are connected directly to the same output of the amplifier, the impedance of the combination of the two speakers becomes 2 Ohm (*not* 8 Ohm), meaning that the load is now double. This connection may result in overloading the amplifier, which could lead to overheating or failure.

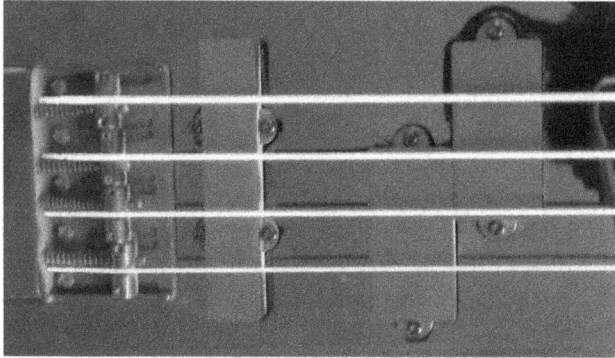

Figure 8.10. Pickups on an electric bass guitar: humbucker near the bridge (left) and two split-coils on the right.

Electric guitar pickup

Special types of microphones, the so-called pickups, are used to amplify the sound of electric guitars. The pickup is similar in principle to the dynamic microphone. A coil is wrapped around a magnet placed just below the metallic string. As the string vibrates, the motion of the metal causes a change in the magnetic field seen by the coil. The changes in the magnetic field induce a voltage on the coil. The change in the voltage has the same frequency as the vibration of the string. This weak signal is sent to the pre-amplifier, as in the case of ordinary microphones. The coils of the pickup are susceptible to the 60 Hz noise from the wall outlets which appears as a 'hum'.

It is common to use double coils of opposite magnetic polarity. The signal from the string has opposite polarity in the two coils, while the 60 Hz noise has the same polarity in the two coils. If the wires at the output of the two coils are cross connected, the signals from the two coils will add and the noise from the two coils will cancel. This cross connection is called '**bucking**' in electronics, and this type of pickup is called the '**humbucker**'. Figure 8.10 shows a humbucker near the bridge (left) and two split-coils on the right, on an electric bass guitar.

8.9 Equations

Ohm's law for AC voltages

For AC circuits Ohm's law reads

$$V_o \sin(2\pi ft) = Z I_o \sin(2\pi ft + \phi)$$

where Z is the impedance connected to the AC voltage source, V_o and I_o are the amplitudes of the voltage and current, respectively, f is the frequency, and ϕ is the phase difference between the current and voltage. For a loudspeaker, $Z = \sqrt{R^2 + (2\pi fL)^2}$, where R is the resistance of the sound coil and L the inductance of the sound coil, measured in units of Henry (H). Note that the impedance increases with increasing frequency, therefore the load gets 'lighter' at higher frequencies.

8.10 Questions

1. Suppose we have a source of constant voltage and two loads of impedance 5 Ohms and 10 Ohms.
   (a) Which load draws more current from the source?
   (b) Which load draws more power?
   (c) Thick short wires allow current to flow freely without impedance. Suppose we connect a thick wire between the terminals of the voltage source. Will this load draw high or low current?

2. (a) Which type of microphone would be more suitable for a singer in a band?
   (b) Which type of microphone would be more suitable for a theater stage?
   (c) Which type of microphone would be more suitable for a news broadcaster in a TV studio?

3. The power gain of an amplifier is 20 dB.
   (a) Find the output power, if the input power is 1 Watt.
   (b) If the efficiency of the amplifier is 50%, how much power must be supplied to run the amplifier?
   (c) Suppose that the output of the amplifier is connected to an 8 Ohm loudspeaker. The efficiency of the loudspeaker is 10%. How much sound power does the loudspeaker put out? What happens to the rest of the power supplied by the amplifier?
   (d) Suppose that we connect a 4 Ohm loudspeaker instead. Will this loudspeaker draw more or less power? Explain.

4. Suppose that we have a non-linear amplifier that has power gain of 100 dB. Explain how the non-linearity would affect the performance of the amplifier under the following conditions:
   (a) When the volume setting is very low.
   (b) When the volume setting is very high.
   (c) Which characteristic of the amplifier (e.g., the gain, the total harmonic distortion, etc.) would most likely be affected by the non-linearity?
   (d) Which frequency ranges (bass, midrange, and treble) would be affected the most by the non-linearity?

5. Assume that an earphone outputs 0.002 Watts into ones ear, and that the ear's opening is $1cm^2$
   (a) How many square centimeters in one square meter?
   (b) Use the result of part (a) to find the intensity of the sound. (see section 1.5)
   (c) How does the above intensity compare to the threshold of pain? (see section 4.3)

Chapter 9

The musical environment

9.1 Introduction

From ancient times, significant thought and effort has been devoted to improve the acoustics of theaters, temples, and other places of public gatherings. A scientific approach to improving acoustics dates back to the late 19th Century, and today, lecture rooms, auditoriums, concert halls, etc, benefit from well-established design principles of sound engineering. But because of the many parameters that need to be factored in, the designs are not always successful. In addition, the quality of the sound in a hall may be best for one particular purpose—for example, an orchestra performance—but unsuitable for singing.

In this chapter, we will apply the concepts introduced earlier to understand the physical quantities that affect the quality of sound, both outdoor and indoor. We discuss which criteria should be applied to make a particular space more suitable for a specific purpose, such as orchestral performances, speech, etc. We also consider how the relevance of the design criteria relates to the size of the space. For instance, what criteria should be used to achieve better sound quality for a music room in a house versus a music hall? We will start with a brief review of the important processes involved in indoor and outdoor acoustics.

9.2 Review of the fundamental processes

The sound emitted from a sound source tends to spread in all directions (see section 2.7). As a result, the intensity of sound decreases as the distance between the source and the listener increases following the inverse square law. So if we have one seat in the front, say at 10 m from a speaker and one seat at 20 m (i.e. we double the distance) and along the same line of sight from the speaker, the intensity at back the seat will be $1/2^2 = 1/4$ of the intensity received at the front seat. Also recall that high frequencies are attenuated by absorption in air more than low frequencies. In addition, low frequencies tend to spread more than high frequencies.

From the reflection of sound (see section 2.4) we know that we can hear our voice reflected from a wall, if our distance to the wall is 17 m or more. So, the round trip distance is $2 \times 17 = 34$ m. As sound travels at a speed of about 343 m s^{-1}, the round trip time is $34/343 = 0.1$ s, or 100 ms. This means that if there is any reflection from a wall that is closer than 17 m (about 50 ft) the sound reflected from that wall will 'fuse' together with the original sound. When sound waves hit an obstacle, the obstacle reflects part of the wave energy and absorbs part of the wave energy. The percentage of absorbed energy depends on the material the obstacle is made of and the texture of the obstacle. Softer materials (drapes, pillows, etc) absorb more (i.e. reflect less) than hard materials (e.g. concrete, hardwood floors, etc).

9.3 Outdoor acoustics

The main concern in an outdoor environment is to minimize the loss due primarily to spreading and absorption. Figure 9.1 illustrates the path of sound waves in an

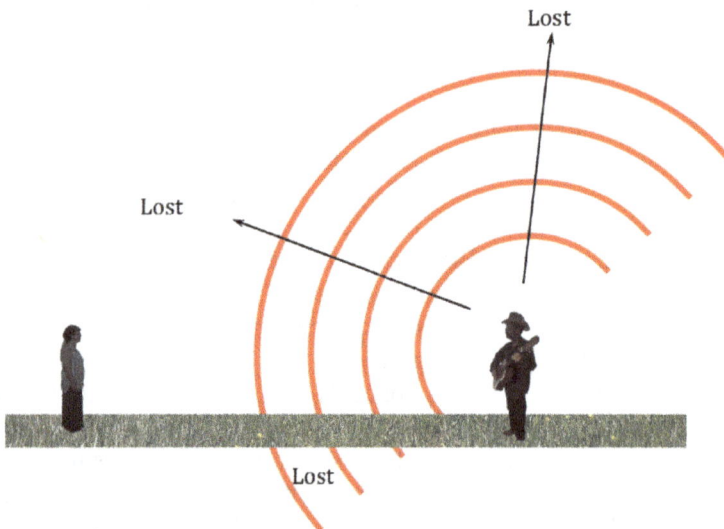

Figure 9.1. Path of sound waves in an outdoor setting.

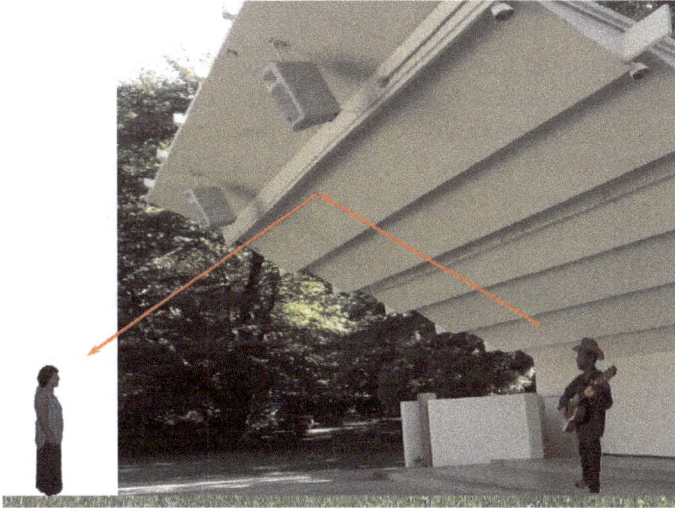

Figure 9.2. A band shell reflects sound towards the audience.

outdoor setting. Note that grass reflects very little energy, while the waves moving upwards do not reach the listener.

One way to remedy the problem is to add barriers behind, above, and to the sides of the stage, as shown in figure 9.2. Properly oriented barriers reflect some of the sound energy towards the audience that would be otherwise lost. Another problem is that if the whole setting is flat, people standing in the front act as barriers that partially block the sound from reaching the audience in the back.

The band shell shown in figure 9.2 partially ameliorates the situation. Lifting the stage would also help, by allowing the audience in the back to have an unobstructed line of sight to the stage, thus receiving more of the direct sound from the stage. If loudspeakers are used, they should be above ground. Note that in figure 9.2, the loudspeakers are mounted at the top of the shell. Another solution is to have the audience area sloping upward from the stage, allowing unobstructed line of sight to the stage. This is the basic idea behind the amphitheater design in ancient outdoor venues, and modern venues.

In large outdoor concert venues, several loudspeakers must be placed at various distances from the stage, to reinforce the sound. In this case, however, listeners in the back hear the sound from the nearest speaker first. The sound from the loudspeakers on the stage has to travel a longer distance to reach them, therefore this sound may arrive noticeably later, like an echo. This can deteriorate the quality of the sound, unless some electronic compensation scheme is used. Delaying the signal going from the amplifier to the distant speakers usually takes care of the lag in arrival times. In other words, the sound from all speakers arrives to the audience in the back approximately at the same time. In outdoor venues, barriers that are just far enough to produce an audible echo (just over 17 m or 50 ft) can be detrimental to the sound quality. This is why good outdoor venues are by design located in open spaces, surrounded by trees.

9.4 Indoor acoustics

As in outdoor venues, the basic objectives of large halls, and music listening rooms in a house, are to get direct sound to the listener and distribute the sound in a uniform way as much as possible. Reflections that reinforce the sound are always present indoors by default. Figure 9.3 shows two such reflections, from the floor and the ceiling. The direct sound reaches the listener first, followed (in this example) by the reflection from the floor. The reflection from the ceiling arrives shortly thereafter, followed by many others. For example, the sound may bounce from the back of the stage and reach the listener, or may bounce from the back of the stage to the ceiling and then to the listener, or from the sidewalls, or the back of the hall, and so on. The combinations are virtually infinite. With each reflection, some sound energy is lost to absorption, but also, to a lesser extent, the air in the hall absorbs part of the energy, especially the high frequencies. So after some time, the sound dies out.

Figure 9.4 shows the evolution of sound received at a typical point in the audience area from a steady tone played at the stage. The first step-like feature marks the arrival of the direct sound from the stage, followed by another step marking the arrival of the first reflection, the second reflection, and so on. As the subsequent reflections arrive with little time lag in between, they fuse into each other, and the power levels off at the **reverberant level**, which is higher than the direct sound. This means that the sound is reinforced by the multitude of reflections.

The combination of direct and reflected sounds together forms the **reverberant sound**. The sound power level remains at the reverberant level for as long as the steady tone is on. After the steady tone is turned off, the power level begins to die out primarily because of absorption at the walls. Figure 9.5 shows how the power level drops. Note that the decay starts at 50 ms. The time it takes for the reverberant sound to drop by 60 dB from the reverberant level (i.e. to drop to one millionth of the reverberant level) is called the **reverberation time**. In figure 9.5 the black line corresponds to reverberation time equal to about 100 ms, and the red line corresponds to reverberation time of about 150 ms. Note that both lines are straight. This is a desirable quality for a music hall.

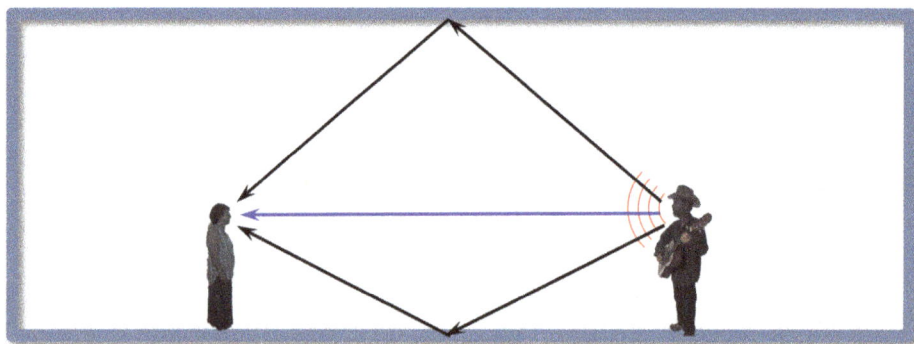

Figure 9.3. Direct sound (blue arrow) and reflected sound (black arrows) in a closed space.

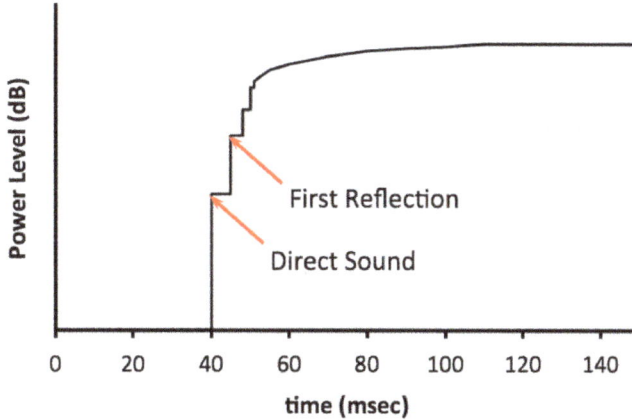

Figure 9.4. Power level for a steady long tone received at a typical point in a hall.

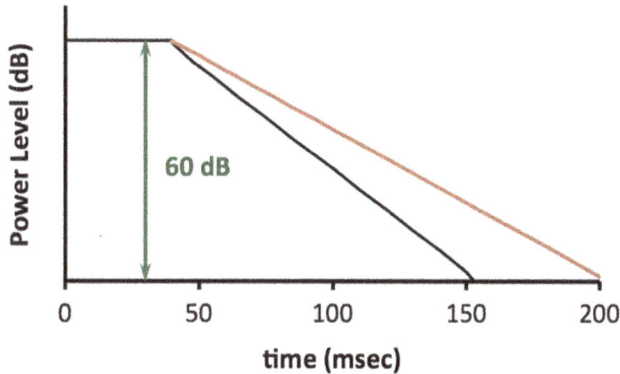

Figure 9.5. Reverberation time for the black line is 100 ms, and for the red line 150 ms.

The major part of absorption occurs at each reflection (by the walls, seats, furniture, audience, etc). In a larger room, the distance between walls on average is longer and the sound waves travel for a longer time between reflections. This means that the sound dies out slower, which makes the reverberation time longer for larger rooms. Also, the reverberation time depends on the percentage of absorption at each reflection. If the walls do not absorb significantly, the sound waves will take many bounces before their energy is lost. In other words, the reverberation time is longer if the absorption at the walls is low, and vice versa. Of course, there are different reflecting surfaces in a room or a hall: walls, chairs, carpets, etc, therefore different percentages of absorption. In addition, the presence of an audience affects the reverberation time as well. When empty, a hall will have longer reverberation time than when it is full of people. The reverberation time also depends on the frequency. Higher frequencies die out quicker than lower frequencies, due to absorption by the air in the room. The calculation of the reverberation time can be rather complicated.

There are specialized room acoustics software that can calculate the reverberation time from the dimensions and the wall materials[1].

9.5 Sound qualities of halls

In describing the sound quality of music halls, terms such as 'fullness', 'intimacy', etc, are commonly used. Here we will relate these terms to measurable quantities. First of all, **liveness** is directly related to the reverberation time. Long reverberation time gives liveness to the hall. Depending on the purpose of the space, a long or short reverberation time may be more suitable. For example, in a lecture hall, reverberation times should be short, about 0.5 s. Longer reverberation times have a negative impact on the clarity of speech. On the other end of the spectrum, to really enjoy organ music, a much longer reverberation time (about 2 s) is more suitable. Symphony halls have reverberation times in the range of 1.5–2.5 s.

Intimacy is used to indicate the feeling of being close to the performers. This occurs when the first reflection arrives less than about 20 ms after the direct sound. In figure 9.4 the first reflection arrives 5 ms after the direct sound, which gives intimacy to the hall represented by the figure. One way to achieve intimacy is to have a reflective surface above the stage, as is shown in figure 9.2. Adding reflecting surfaces on the sides helps as well, and in addition it helps blend the sound. **Blend** is important for large stages, otherwise at some spots in the hall some voices/instruments may sound louder/weaker than intended.

If the direct sound is a small portion of the reverberant sound level then we perceive mostly the reinforcing reflections. A long reverberation time combined with a reverberant level that is much higher than the direct sound result in what is known as **fullness**. The opposite of fullness is **clarity**, which results when the reverberation time is short and the reverberant level is not much larger than the direct sound (i.e. the reinforcement from the reflections is weak).

If the reverberation time for low frequencies is longer than the reverberation time for high frequencies then we speak of **warmth**. In the opposite case, i.e. if the reverberation time is about the same for all frequencies, then we speak of **brilliance**. Again, these two characteristics are in a way opposite, meaning that some compromise in the design may be necessary depending on the main purpose of the hall.

9.6 Sound qualities of small rooms

The walls in most rooms in homes are very close to the listener, which means that the first reflection arrives well under 20 ms after the direct sound. In other words, intimacy is always there. If the average dimension of the room is 5 m (15 ft) then sound bounces off the walls over 70 times in each second, which means the effect of absorption will be very significant compared to a large hall. So, because of their small size, and the relatively high absorption on surfaces usually found in a room

---

[1] For further discussion see section 9.12. A free online calculator can be found at: https://www.atsacoustics.com/page–Free-Online-Room-Acoustics-Analysis–ora.html.

(carpets, drapes, etc), the reverberation time is usually less than 0.5 s. Short reverberation time means that liveness is always lacking in home listening rooms. The multitude of reflections in a small room reinforce the sound, in other words the reverberant level is higher than the direct sound. But as the reverberation time is so short (remember that the second requirement for fullness is long reverberation time) home listening rooms lack fullness. Therefore, the fullness must be captured during recording in a concert hall or added electronically in the recording studio.

9.7 High fidelity sound

To enjoy high quality recorded or transmitted sound, the reproduced sound must correctly include all the frequencies present in the original performance, without noise or distortion of the loudness, or the range of loudness[2]. All these requirements are met by most home sound systems. In addition, the reproduced sound should capture the spatial pattern and reverberation characteristics of the original sound. If the original performance had the violins on the left and the cellos on the right of the audience, then the listener at home should also get the sensation of this spatial distribution of the instruments. This requires at least two tracks of sound, and two or more loudspeakers. A system that meets all these requirements is a **high fidelity (HiFi)** system. **Stereophonic** and **surround sound** systems meet these two requirements with varying degrees of success.

9.8 Stereophonic sound

Our binaural hearing (see section 4.12) allows us to identify the direction of the sound source. Our brain uses both the difference in arrival time and the difference in loudness to localize the sound source. As discussed above, the audience in a concert hall receives the direct sound first followed by the reverberant sound, which comes from all directions and can be louder than the direct sound. Under these conditions the audience can identify the direction of the sound as coming from the orchestra because of the **precedence effect.** The precedence effect tells us that our hearing is tuned to identify the direction of the first arriving signal as the source of the sound.

On the other hand, in a small room setting if two identical sounds arrive from two different directions within 2 ms or less, then the source is perceived as being somewhere in between the two actual sources of the sound. If the intensity of one of the two sounds is higher, then the source is perceived as being somewhere in between the two sources, and closer to the side of the source with higher intensity. If the distance from the listener to the loudspeaker changes by 1 m (about 3 ft) the arrival time changes by about 3 ms. Therefore, one can appreciate that the placement of the loudspeakers relative to the listener is very important in reproducing the spatial distribution of the instruments.

In the simplest form, a stereo recording reproduction starts by placing two microphones: one picking up mostly the sounds from the left side of the stage (L) and one for the right side (R) of the stage as shown in figure 9.6(a). The sound picked up by each

---

[2] The range of loudness, or dynamic range, is discussed in the concluding part of section 8.6.

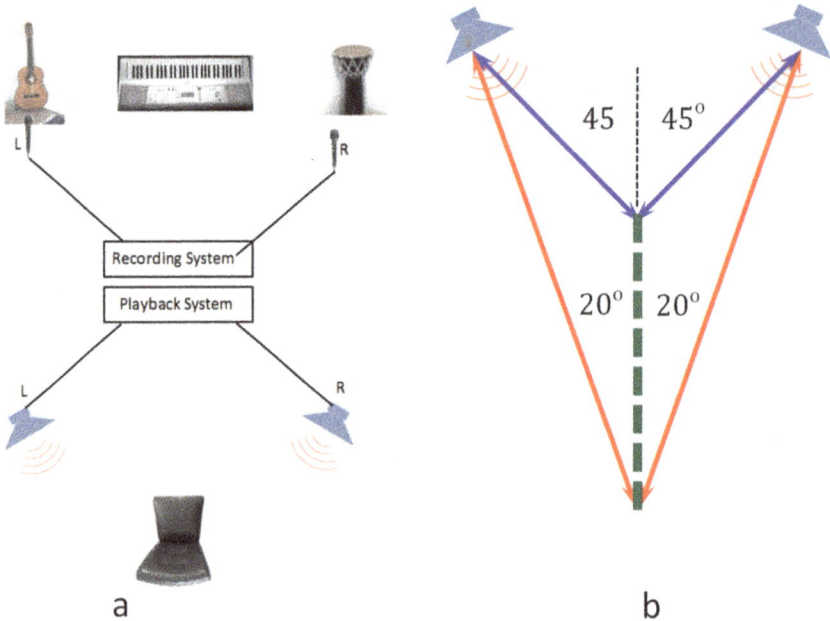

Figure 9.6. (a) Stereo recording and reproduction. (b) Placement of loudspeakers for stereo systems.

microphone is recorded on a separate channel, L and R, respectively. On playback, the sound from each recorded channel is directed to a separate loudspeaker, L and R.

Of course, the sound from each loudspeaker will bounce around the room, and the reflections will be coming to the listener from everywhere. However, if the listener is seated midway between the two loudspeakers, the direct sound from loudspeaker L will come to the listener first and at higher intensity, hence the listener will identify loudspeaker L as the source. So, what was recorded by microphone L will appear as coming from the loudspeaker located on the left side. Similarly, the sound recorded by microphone R will appear as coming from the speaker on the right hand side. If there are sounds that are picked up equally by microphones L and R (presumably an instrument in the middle of the stage) and reproduced equally by loudspeakers L and R, then as long as the listener is seated midway between L and R, the sound will come through both loudspeakers at the same level and at the same time. Therefore, this sound will appear as coming from the middle.

9.9 Placement of loudspeakers

Stereo sound relies on the arrival time and intensity of the signal received from the two channels, therefore configuration of the listening space is important. Figure 9.6(b) shows a typical arrangement, where the speakers are placed near the corners of the room, and along the shorter wall. One reason for placing the speakers near the corners is that the sound is reinforced by reflection from both walls forming the corner. The reason for using the shorter wall is that it allows for more 'good listening space', the so-called **sweet spots**. This can be understood by examining the angles indicated in figure 9.6(b). For best results the listener should be seated on the center line between the speakers, indicated by the green dashed line. The speakers should be at an angle of about 20–45° to the right and left of the listener. If the angle is too small, the stereo effect is diminished (we essentially hear one source, i.e. a **monophonic** sound) and if the angle is too large we perceive two different sources with a 'hole' in the middle. In figure 9.6(b) the red solid lines indicate the 20° angles, and the blue lines are the 45° angles. The listener can be seated anywhere along the green line and have good stereo effect. If we assume that the speakers are set 4 m apart (12 ft) then the distance of the listener to the speakers must be 2.8 m (8.5 ft) for the 45° angles, and 5.8 m (17 ft) for the 20° angles. If the speakers were set 6 m (18 ft, which implies a wider wall) then the distances of the speakers to the listener should be 4.2 m (12.7 ft) and 8.5 m (26 ft) for 45° and 20° angles, respectively. These distances are too large for an ordinary house, and leave little room for placing the furniture without obstructing the line of sight from the listener to the speakers.

As mentioned, source localization depends on the intensity and time of arrival, therefore one can use the **balance** control in a stereo system to adjust the sweet spot. In other words, the listener can sit to the right or left of the green center line, and increase the loudness of the left or right channel, respectively, and still have good stereophonic sound. The balance adjustment works as long as the difference in arrival times remains under 1–2 ms; in other words, one can move at most about 2 m (6 ft) right or left of the center line and adjust the balance to restore the stereo effect.

9.10 Ambient noise

Another factor that affects the quality of sound is noise. For external noise, a design that includes panels with the correct noise reduction coefficient (NRC) is important. The NRC is a number that tells us the fraction of the noise absorbed. For example, NRC = 0.1 means that 10% of the incident sound is absorbed and 90% is reflected back; NRC = 0.8 means that 80% of the sound is absorbed and 20% is reflected. In addition to external noise, ventilation and air-conditioning systems can be a major contributor of noise, both in large halls and small rooms. The noise is essentially guided into the space by the ductwork, especially metal ductwork, and is particularly noticeable around the registers.

9.11 Further discussion

Arranging microphones for stereophonic recording

There are different ways of arranging the microphones for stereo recording. A common method is to have two omnidirectional (or cardioid) microphones spaced 3 m (about 10 ft) apart (see section 8.4). The pair must be located in front of the stage, and at a distance comparable to the lateral extent of the group performing on the stage. Another method uses two cardioid microphones on the stage, separated by a small distance, 30 cm or less (less than 1 ft), with their axes of maximum pickup (the 0° line in chapter 8, figure 8.5(b)) forming an angle of about 90–100°. One of the microphones (either one) is pointing to the right half of the performing group and the other to the left.

Surround sound

A surround sound system adds three (5.1 system), four (6.1 system), or five (7.1 system) loudspeakers plus a subwoofer (the .1 indicates subwoofer) to the L and R of the stereophonic system of figure 9.6(b) (see section 8.7 for discussion of loudspeakers). Depending on the number of additional speakers, one can be located front-center, two on the sides, and two behind the listener, providing the 'surround' sensation. The signals to the additional speakers are a combination of the L and R sounds, or can be used exclusively for sound effects, if the system is used for video. The signal to the additional speakers is electronically delayed compared to the sound of L and R, to optimize the surround effect at the point of the listener.

Resonance in indoor spaces

In section 6.9 we found that the frequency modes for a pipe closed at both ends are

frequency = $(n \times 343)/(2 \times \text{length of pipe})$

where $n = 1,2,3, \dots$ etc. We can assume for simplicity that a closed room or hall has a cylindrical shape. Let us assume that the closed pipe has length equal to the long dimension of the room, say 6.86 m (about 20 ft), and calculate the first few frequencies that can build up. Using the above equation we find

25, 50, 75, 100, 125, 150 Hz, and so on.

Note that the frequencies are spaced apart by 25 Hz, i.e. the spacing of the overtones is equal to the fundamental frequency. In a large concert hall, say 50 m long (150 ft) the fundamental frequency is 3.4 Hz, therefore, the overtones are spaced apart by 3.4 Hz, which is much smaller than the spacing of the frequency modes in a small room. Of course, the hall has sidewalls, floor, and ceiling, which add more frequency modes. The result is that the spacing of the frequency modes in a large hall is so small that we essentially end up with a continuum of frequency modes. This is not so for a small room.

Figure 9.7 shows the first five modes in a room 5 m × 6 m × 3 m (15 × 18 × 9 ft). We note that the spacing of the frequency modes remains significant, even when we include reflections from all sides of the room. Keep in mind that if the reflections can

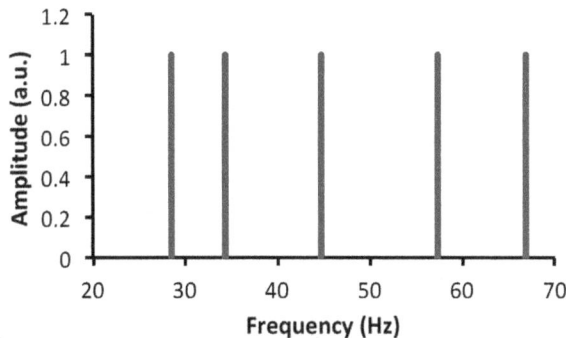

Figure 9.7. First five modes in a room 5 m × 6 m × 3 m. Amplitude is in arbitrary units (a.u.).

build up a standing wave of some frequency, then this frequency is also a resonance frequency (see section 6.12), meaning that in small rooms some discrete frequencies will resonate, and the effect may be quite noticeable. Experimenting with the orientation of the speakers or the furniture may be helpful in this situation.

From the discussion of closed pipes (see section 6.9), we know that the walls represent antinodes for the pressure, i.e. the oscillation of the pressure is the maximum. This is true of small rooms as well. The air pressure can oscillate the maximum at the walls. Alternatively, if we want the loudspeakers to set the pressure oscillation in the room efficiently, placing them near the walls would be the best spot. This is another reason for placing the speakers near the corners of the room.

9.12 Equations

The Sabine equation

The reverberation time can be calculated from the Sabine equation:

reverberation time = 0.161 (volume of room/total absorption).

Each material has its characteristic absorption coefficient. Therefore, to find the total absorption by the surfaces in a room we need to consider all the surface materials separately. For example, we will consider an *empty* room with dimensions $5 \times 6 \times 3$ (width $\times$ length $\times$ height, all in meters). The volume is $5 \times 6 \times 3 = 90$ m^3.

We will assume hardwood floor, drywall, and acoustic tile ceiling. The surface areas (with the typical absorption coefficients in parentheses) are:
- Floor $5 \times 6 = 30$ m^2 (absorption coefficient = 0.3).
- Ceiling $5 \times 6 = 30$ m^2 (absorption coefficient = 0.7).
- Two side walls of $5 \times 3 = 15$ plus two side walls of $6 \times 3 = 18$ or 66 m^2 total (absorption coefficient = 0.1).

To find the total absorption, we multiply the total surface for each material times the absorption coefficient of the material, and add. We find[3]

$0.3 \times 30 + 0.7 \times 30 + 0.1 \times 66 = 36.6$ m^2 or **sabins**.

Using the above values in the Sabine equation we find

reverberation time = $0.161 \times (90/36.6) = 0.4$ s or 400 ms.

Note that in the above example we used meters to measure the dimensions of the room. The calculation is the same for measuring in feet, except that the coefficient 0.161 in the Sabine equation should be changed to 0.05 instead. Note also that the absorption coefficient values used are 'typical'. The absorption coefficient depends on the frequency as well, hence the dependence of the reverberation time on the frequency as discussed in section 9.4 above.

---

[3] The unit **sabin** is commonly used, in honor of Wallace Sabine (1868–1919) who pioneered concert hall acoustics.

9.13 Questions

1. A small lecture hall is 10 m long. Here we will ignore the reflections from the sidewalls, ceiling etc.
    (a) How long does it take for the direct sound to reach a student sitting in the middle of the hall?
    (b) How long it takes for the sound to complete 1 round trip between front and back walls?
    (c) Suppose the walls absorb 50% of the incident sound. What fraction of the sound energy is lost after 1 round-trip?
    (d) Suppose that the walls absorb 10% of the incident sound. What fraction of the sound energy is lost after 1 round-trip?
    (e) Which wall material would make the reverberation time of the hall longer?
2. A concert hall is 34 m long. Here we will ignore the reflections from the sidewalls, ceiling etc.
    (a) How long does it take for the sound to complete 1 round trip between front and back walls?
    (b) Suppose that the walls absorb 10% of the incident sound. What fraction of the sound energy is lost after 1 round-trip?
    (c) Other things being equal, will a larger hall have longer or shorter reverberation time?
    (d) Is the sound in this hall more likely to have 'intimacy' compared to the smaller hall of problem 1? Explain your answer.

Chapter 10

Sound recording and reproduction

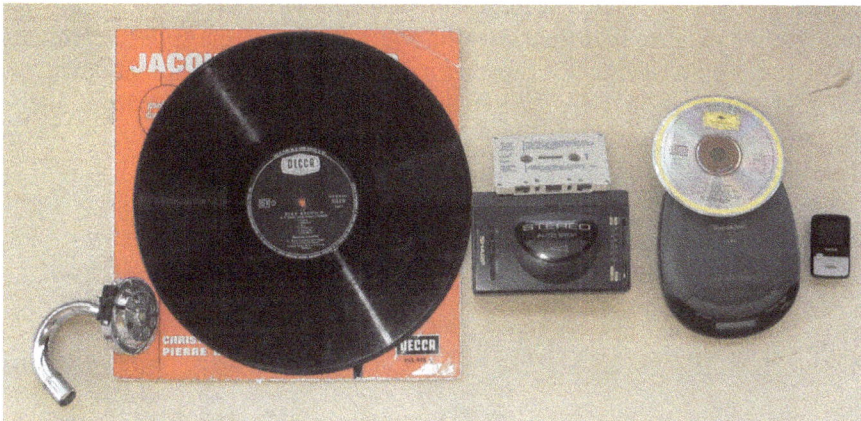

10.1 Introduction

Commercial recording and reproduction of sound dates back to the late 1800s when Thomas Edison patented his phonograph. The early models could both record and playback using rubber based or wax cylinders. The disk format was introduced in the early 1900s, and used for playback on gramophones. The gramophone disks were made of shellac originally and then from a polyvinyl chloride base, hence the term 'vinyl'. The early gramophones used spring driven motors, which required frequent cranking. By 1940 fully electrical units with electronic amplifiers replaced the older designs.

While records are still in production, and for some people remain the favorite medium, other recording and playback innovations have become popular as well. The most successful invention used magnetic media (magnetic wire, then magnetic tape). In the early 1980s, CDs were commercialized, and quickly became the preferred medium for about two decades. In this chapter we will describe the basic principles of some analog and digital recording and reproduction systems. Some of the concepts used here, like induction, frequency bandwidth, distortion, etc, were introduced in chapter 8.

10.2 Gramophones

In a gramophone record the signal is recorded on a vinyl disk. The disks have grooves spiraling from the outer edge towards the center. The walls of the grooves make an angle of about 45 degrees to the vertical, and the sound signal is impressed as ridges on the walls, as shown in figure 10.1.

In a stereo system, there are two channels of signal, one on each wall of the groove. The sound signal to be recorded is converted into an electrical signal, which sets a **cutting stylus** into vibration. The vibrations of the stylus are engraved on a coat of shellac or lacquer. The grooved surface is then coated with a metal film, so the metal film becomes a 'negative' of the groove pattern. The metal negative is used as a print, to impress grooves on a vinyl record. For playback, a **stylus** is in contact with the rotating groove pattern, and the ridges on the walls make the stylus move laterally (right–left) as well as vertically (up–down). In the early models, the stylus was a 'needle' and the vibrations were fed to a diaphragm that produced audible sound with a horn (see opening figure of chapter 9) or without a horn. In some models the sound was heard using a stethoscope, i.e. a precursor of today's earphones.

The record is placed on a rotating **platter** (or **turntable**, although some use this term to refer to the whole gramophone unit). The platter is usually heavy (1–3 kg, or 2–6 lb) to minimize fluctuations in the rotation speed. The platter is set to rotation

Figure 10.1. Microscope image of the groove in a gramophone record. Image credit: (https://commons. wikimedia.org/wiki/File%3ALangspielplatte.jpg).

by an electric motor. In one design, the motion of the motor's shaft is transmitted to the platter by a belt (**belt drive**). In another design, the shaft of the motor is directly attached to the axis of the platter (**direct drive**).

10.3 Magnetic tape deck

Recording sound signals on magnetic material was commercialized in the mid-1940s and became very popular with the introduction of the compact cassette in the early 1960s. The method is based on the interrelation between electric signals and magnetic fields, which was discussed in section 8.2, and the interaction between magnets. In a cassette the plastic tape is coated with finely powdered magnetic material such as iron oxides or chrome dioxide. These tiny magnets can be aligned by the action of a magnetic field, in the same way that the compass needle aligns in the direction of the Earth's magnetic field. Once aligned, the particles retain their alignment, even after the external aligning field is removed.

Figure 10.2 schematically illustrates the structure of a recording/playback **head**. The signal to be recorded is fed to a coil (red) wound on an iron ring (blue). The current in the coil produces an analogous magnetic field (green lines) in the ring. A small gap in the ring filled with non-magnetic material (gold) 'kicks out' some of the magnetic field from the ring towards the tape. This field aligns the tiny magnets on the tape (brown), which is moving past the head at constant speed. The pattern of alignment of the magnetic powder follows the pattern of the input signal. In playback the process is basically reversed: as the tape is run past the head, the magnetic field from the powder in the tape produces a magnetic field in the ring, which in turn induces a voltage in the coil. This voltage is fed to the amplifier. Some units use the same head for recording and playback. Better units have separate heads for recording and playback. In these units the recording head has a wider gap (about 4 microns) than the playback head

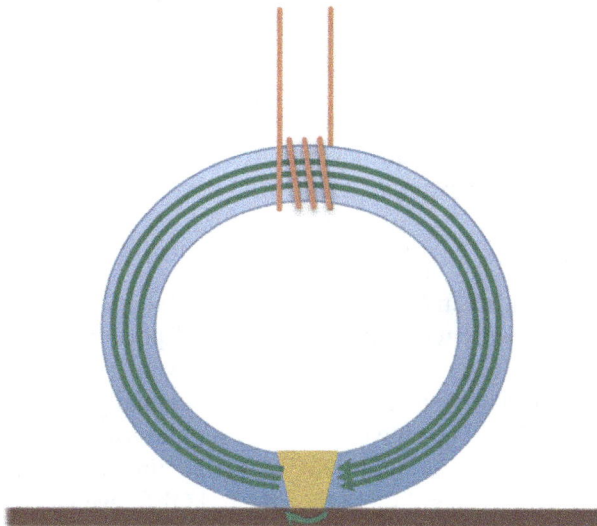

Figure 10.2. Structure of a recording/playback head of a tape player.

Figure 10.3. A tape head (right) and motor driven capstan system and rubber wheel (left).

(1 micron)[1]. In addition, an erase head is used to make the alignment of the particles random before recording again. Random alignment means that the sum total of the magnetic field of the particles over a distance equal to the gap of the head is zero. The tape is moved past the head using a system of a motor driven **capstan**, and a rubber wheel pressing against the capstan, shown in figure 10.3. The tape lies flat between the rubber wheel and capstan. It is important that the tape remains flat against the head and move at constant speed. Therefore, some expensive units have two additional motors to keep the tape tight. For stereo sound, the head must have two tracks, one for each channel.

The magnetic pattern recorded on the tape is analogous to the waveform representation (see section 3.3). If the image is more spread out horizontally, one can see the oscillations more clearly. If the tape moves fast, the pattern is more spread out, and the performance at high frequencies is improved. Professional units use speeds of 38 cm s^{-1} (15 inch s^{-1}) or higher.

Tape decks suffer from noise (hissing), which comes mostly from the fact that the magnetic particles are finite in size. Even if the tape is erased, i.e. the magnetic field of the particles points at random directions, the playback head reads each individual particle and the result is a high frequency hiss. Most commonly, the decks come with a noise reduction system—for example, DBX or Dolby—to correct the noise problem.

10.4 Compact disc

The compact disc (CD) became commercially available in 1982, and its popularity increased rapidly, especially with the introduction of portable CD players in 1984. The first CDs were read-only. Recordable CDs were released in 1990, which increased their popularity even more. The CD essentially reflects light, and the amount of the reflected light can have two levels: high or low. As discussed in section 7.2, this two-state recording is essentially a digital binary signal, consisting of a sequence of 0s and 1s corresponding to the ON and OFF levels of the reflected light.

---

[1] 1 micron is one millionth of a meter.

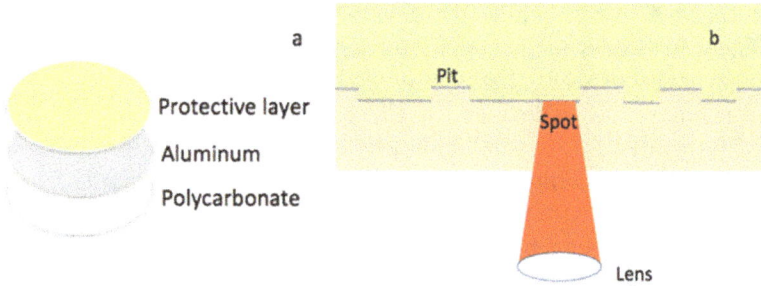

Figure 10.4. (a). CD layers. (b) Laser beam focuses on the pit pattern.

The digital signal is impressed on a clear polycarbonate surface, as a series of **pits** of uniform depth. The pitted surface is coated with layer of reflecting material (usually aluminum), and another layer of protecting coating is applied on top of the aluminum. The sequence of layers and the pit pattern are illustrated in figure 10.4.

To read the signal, a laser beam (red in figure 10.4(b)) coming from the bottom of the polycarbonate is focused on the reflective aluminum layer. If the spot of the laser beam is over a flat part of the pattern, the amount of reflected light stays the same. This is the binary 0 state. But when the spot crosses an edge of a pit the amount of reflected light drops (see section 10.6 for more details). This is the binary 1 state. The reflected light is picked up by a system of light detectors (called **photodiodes**), and is interpreted and transformed to an audible analog signal, as discussed in section 7.3. The pits are impressed in an outwardly spiraling pattern, much like the vinyl records. A motor turns the CD, and a separate mechanical system keeps the laser and the photodiodes that read the reflections on track.

Note that in this scheme there is no physical contact between the device that reads the information and the surface that contains the information; in other words, the CD does not wear out with use, as is the case with records and tapes. Another advantage is that the signal is binary, which means that noise from other components of the system does not affect the reading of the signal.

Typically, a CD will store about 70 min of music. The code used by the manufacturer, the so-called CDDA, can be read by a CD player or a computer disc drive. The contents of a CD can be 'ripped' by computer software, and converted to a file with tracks and information. Microsoft uses the 'wav' format (indicated as '.wav' in the file name extension) and Apple uses the 'aif' format ('.aif' extension). The popular MP3 format, takes up less memory space by using a sophisticated scheme that removes some of the sound frequencies. Recall from section 4.9 that if two frequencies are close enough, one frequency could mask the other. Therefore, some masked frequencies can be removed without affecting the sound quality noticeably.

10.5 Semiconductor storage devices

Advances in computer storage technologies led to the development of semiconductor storage devices, with storage capabilities far exceeding those of CDs. An added advantage is that semiconductor storage devices have no moving parts, therefore no motors are needed. Additionally, the devices are much smaller, lighter, and more durable. The transistor is the heart of these devices. Recall from section 7.3 that the transistor can act as a switch, therefore it can have two states (ON or OFF) that can be used for binary logic devices. In the case of memory devices, millions of transistors[2] integrated into a small chip are used for storing huge amounts of data (billions of bytes). The storage is 'permanent' (as opposed to 'volatile'), i.e. the data remain in the memory without need of constant refreshing, which would require some electrical power, like a battery. Permanent does not imply that the data cannot be erased. In fact, the data stored in a memory card can be erased and re-written, and the data is not limited to audio, as the card can store all kinds of files, including video, pictures, etc. The memory card can be used either as secondary storage on a device (e.g. flash or thumb drive on a computer, tablet, etc) or it can be packed with a small playback device, as in MP3 players. In this case of course, a battery is required to run the player.

In summary, semiconductor memory storage has all the advantages offered by the CD with the additional advantages of more memory and smaller size. As a result, semiconductor memory has gained great popularity as *the* audio storage medium, while the other media (records, tapes, and CDs) have declined.

10.6 Further discussion

Record player cartridge

In modern units, the stylus is attached to the **cartridge**, shown in figure 10.5(a), which in most cases is similar in principle to the dynamic microphone. In one design, the stylus end (opposite the one contacting the record's surface) has a small coil attached. The coil moves in a magnetic field, thus producing an electric signal. This design is the **moving coil** (MC) cartridge. In another design, the magnet is attached to the stylus, and the magnet moves inside a coil. This is the **moving magnet** (MM) cartridge. The cartridge is attached to the **tone arm**, which has a delicate balancing

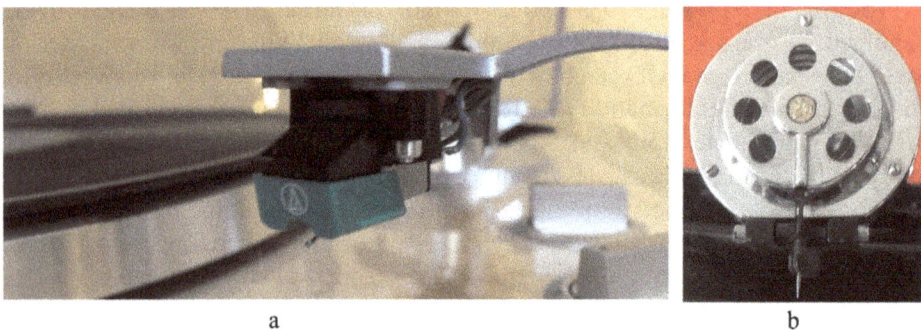

| a | b |

Figure 10.5. (a) Gemini cartridge on TEAC turntable. (b) Cartridge of 1930s 'His Master's Voice' Suitcase Phonograph. The needle is at bottom center.

---

[2] The metal-oxide semiconductor field emission transistor (MOSFET) is the most common at this time.

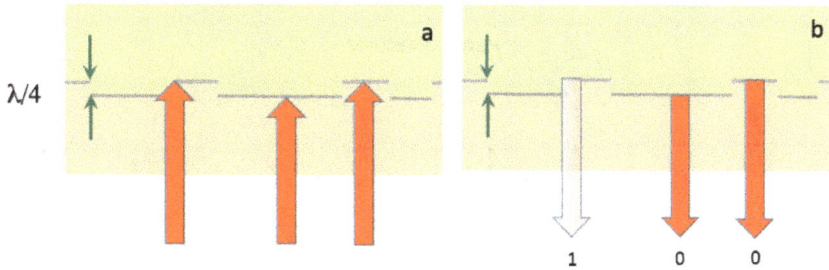

Figure 10.6. (a) Laser spot incident on a CD at three different instances. (b) Light reflected from the aluminum layer. The weak reflection (pink) corresponds to binary 1 and the strong reflections (red) correspond to binary 0.

mechanism that adjusts the force exerted by the stylus on the surface of the record, typically less than 5 g (0.18 oz). Figure 10.5(b) shows the cartridge of an older unit.

Reading data from a CD

The depth of the pits in a CD is selected to be equal to one quarter of the wavelength (λ) of the laser light used by the CD player. Figure 10.6(a) shows the light spot at three different instances. When the laser beam hits the edge of a pit, part of the light reflects from the bottom of the pit, and part of the light reflects from the top of the pit. In other words, part of the light has traveled an extra distance of $\lambda/2$, where λ is the wavelength of the light in the polycarbonate material of the CD. Recall that one half of a wavelength means that the two parts of the reflected beam are 180° out of phase, therefore some cancellation occurs as the light spot crosses an edge. The pink arrow in figure 10.6(b) indicates this reflection. Cancellation does not occur when the spot hits the flat parts at the top or bottom of the pits, as indicated by the red arrows in figure 10.6(b).

10.7 Questions

1. List two advantages of magnetic tape recording over vinyl disk recording.
2. List two advantages of CD over magnetic tape recording.
3. List two advantages of semiconductor storage devices over CDs.
4. We could increase the storage capacity of CD (and vinyl disks or tapes) by rotating the CD slower. If we rotate 100 times slower, we would store 100 times more songs! Obviously there is a limit and here is a very rough explanation. Suppose that we have sinusoidal pure tone of 10 000 Hz, and assume that each edge in figure 10.4b is a positive peak of the sinusoidal.
   (a) How many positive peaks does the sinusoidal have in 1 second?
   (b) How many edges do we need to record the waveform for 1 second?
   (c) Assume that on average one complete rotation of the CD we can fit 5000 equally spaced edges. How many full rotations are needed to fit 10 000 edges?
   (d) How many full turns do we need to read-out in 1 second?

Chapter 11

Musical instruments

11.1 Introduction

The use of musical instruments dates back to the Paleolithic age. Archeologists have found flutes constructed about 40 000 years ago from animal bones. Other instruments made from materials that disintegrate, such as drums made of animal skin, may have been in use at that time as well. Many musical instruments evolved from early forms that date back to ancient China, Egypt, etc. Instruments such as the piano were invented in the past few centuries, and have no ancestors in antiquity. The development of musical instruments, the shape, and the selection of the materials, was based on intuitive understanding of the properties of sound, and the process of trial and error. Yet, the sound quality of some of the instruments constructed this way remains unsurpassed; for example, the famous Stradivarius violins, made between approximately 1666 and 1737 AD by the Stradivari family in Cremona, Italy. Scientific understanding of the properties of sound and the properties of materials gave us detailed insight on musical instruments, and in many cases led to improvements of the sound quality of the instruments. In this

doi:10.1088/978-1-6817-4680-7ch11　　　11-1

chapter we discuss the basic types of musical instruments, and the principles underlying their design.

11.2 Basic functions and types of musical instruments

A musical instrument essentially converts energy supplied by the player into sound waves with characteristics that to a large extent are controllable by the player. The basic characteristics of the tones produced are pitch, loudness, duration, and timbre. The degree of control allowed depends on the nature of the instrument. For example, a pipe organ does not allow much control of the loudness, and crash cymbals allow very little control of the pitch. Similarly, the loudness of the tones produced by a recorder can be increased to some extent by blowing harder, up to a certain point, beyond which blowing harder causes the pitch to jump, resulting in a different tone altogether.

One of the basic functions of a musical instrument is to produce tones of the desired pitch. This is achieved by creating standing waves in a clamped string, or an air column, or a stretched membrane, etc. Let us consider a guitar for example. Plucking a string of a guitar creates a tone. So, the **excitation** occurs by plucking. Plucking imparts energy to the string, which begins to vibrate. As discussed in section 6.2, the vibration occurs only at certain frequencies. In other words, we have excited a **vibrator** (the string) of a certain pitch. The vibrating string by itself does not transfer its energy to the surrounding air efficiently, and the loudness of the tone is very low. Therefore, it is essential to transfer the energy from the vibrating string (the vibrator) to an efficient **radiator**, which is the body of the guitar.

The string and air column vibrators discussed in chapter 6 were ideal, and produced a number of harmonic overtones, i.e. mode frequencies that were integer multiples of a fundamental frequency. In the case of strings, this implied that the string was thin and perfectly flexible, which we will also assume in the case of two-dimensional plates and membranes (see section 11.5). In the case of air columns, we assumed that the pipes were straight, had a perfectly cylindrical bore, and that the pressure nodes were exactly at the open end of the cylinder. These assumptions do not apply exactly to real materials and instruments. For example, all strings have some stiffness, and trumpets are far from straight pipes. Therefore, some deviations from our basic conclusions are to be expected when discussing real instruments. The most significant deviation is that the frequencies of the overtones are not integer multiples of a fundamental. Instead, string overtones become **anharmonic** (see section 6.14) because the stiffness increases the frequency of the higher overtones. Similarly, the overtones of a trumpet are anharmonic because the trumpet is not a straight pipe.

Depending on the type of vibrator, most of the musical instruments can be classified under three categories. In **wind instruments** the vibrator is an air column, as in a flute. **String instruments**, such as the violin, use the vibrations of strings. In **percussion instruments**, the vibrator is a rod, as in chimes, or a surface like the membrane of a drumhead. This classification is based essentially on the sections of a symphony orchestra, and certainly does not cover all instruments. Depending on other specifics, musical instruments are classified into subcategories. A wind instrument can be classified as a reed instrument (e.g. the clarinet) or brass instrument (e.g. the trumpet). In the following sections we discuss the basic characteristics of each type, with selected examples.

11.3 Wind instruments

Wind instruments are essentially based on forming standing waves in the air column inside the **bore** of the instrument. Wind instruments have much more complicated shapes than the straight cylindrical pipes discussed in chapter 6. For example, the bore of some wind instruments like the flute, is cylindrical, but has holes on the side of the pipe. Others have a conical bore, like the clarinet. Others, like the trumpet, have a cylindrical bore that loops around, and transitions into a flared bell-shaped open end. As a result of the intricate shape of the air column, a wide range of fundamental tones and harmonics are allowed, which make these instruments richer. Although the conclusions of section 6.5 may not strictly apply to real wind instruments, pipes are useful as a model to better understand wind instruments. For example, longer air columns (like the tuba) produce lower frequencies compared to short air columns (like the flute).

It is important to recall that the mode frequencies of a pipe (or a string for that matter) are also the resonance frequencies of the pipe (or string). In other words, if we manage to introduce pressure waves of a wide range of frequencies, only the frequencies that match the mode frequencies will build up standing waves, and only these frequencies will resonate, i.e. eventually produce a tone. Therefore, exciting a standing wave in a wind instrument can be thought of as continuously injecting bursts of air, which is essentially noise. Noise contains all frequencies, as discussed in section 5.10. Of all the frequencies injected, the only ones that survive and resonate will be the ones that match the frequency modes of the bore. A simple example is what happens when we blow sideways across the mouth of a bottle; we are blowing-in noise, i.e. a continuum of frequencies, and only the frequency that matches a mode of the bottle resonates. We can think of this process as *channeling* the energy of the air-stream we blow in, into the frequencies that resonate. Wind instruments use intricate excitation mechanisms that allow more control over the characteristics of the sound produced. The excitation mechanisms used in connection with wind instruments operate as **valves**[1] that can repeatedly open and block the airflow into the air column. So, the air entering the instrument is a sequence of 'puffs' that have no discernible tone.

There are three ways to generate these puffs. One way is to blow onto a flexible **reed** and thus set the reed into vibration. The reed is firmly attached to one end of pipe, the **mouthpiece**, leaving a small gap that allows air to flow into the instrument. The structure of the mouthpiece is shown schematically in figure 11.1.

The player's lips seal around the mouthpiece (and reed) near the right end of the figure, blowing a continuous airstream. As the airstream flows through the narrow

---

[1] Not to be confused with the valves of some brass instruments that will be discussed below.

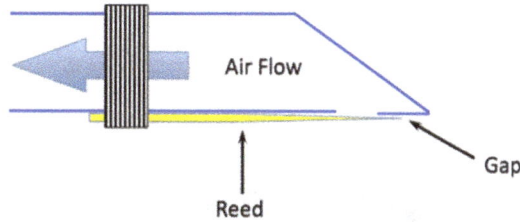

Figure 11.1. Structure of a reed mouthpiece.

gap the air moves faster and creates an upward suction on the reed[2]. If the reed is flexible enough, it will bend upward and seal the gap, and stop the airflow into the bore. The pressure at the gap at this point is low, and the gap will remain closed until some high pressure builds above the reed and forces it open again. In other words, the mouthpiece end acts as a closed end of a pipe. The puff of air that entered through the gap consists of waves of many frequencies. When the waves reach the open end of the bore, as discussed in section 2.4, they will reflect back into the air column and travel towards the gap. After a few round-trips, the only frequencies that will survive are those that correspond to a pressure node at the open end and a pressure antinode at the gap. When the antinode at the gap reaches a pressure 'high' the pressure will push the reed down and open the gap. Once the gap is open, the airstream flows through the gap again starting a new cycle, and eventually a stable tone is established.

It is important to note that the 'open–close' cycle of the 'valve' action of the reed is controlled by the standing wave created in the pipe; in other words, the reed follows the resonant frequencies of the air column. This is an example of **feedback**, where the excitation (valve action of reed) is fed into a resonator (air column) and the wave selected by the resonator is sent back to the reed in a way that controls the timing of the excitation (valve action of the reed). In our example, the standing waves have maximum pressure variation (antinode) at the gap. According to section 6.5 we have a semi-closed pipe. Therefore, wind instruments that use reeds act as semi-closed pipes. Reed instruments include the **clarinet** and **saxophone**. Some instruments use double reeds, based on the same principle described above. Double reed instruments include the **bassoon** and the **oboe**.

In brass instruments, like the **trombone**, the **trumpet**, and the **tuba**, the lips that are pressed closed against the cupped end of the mouthpiece do the 'valve' action. As the player blows, the sealed lips temporarily open and a puff of air is fed into the bore. This puff consists of waves of many frequencies, which are reflected back at the open end. After a few round-trips, the frequencies that correspond to a pressure node at the open end and an antinode at the mouthpiece will establish a standing wave. When the pressure variation at the mouthpiece reaches a 'high' it pushes the lips open and another puff comes in the air column, and the cycle repeats. The result is that the rate at which the lips open and close is controlled by the frequency of the standing waves in the air column, i.e. a feedback loop is established as in the reed

---

[2] The creation of a pressure low by a fast moving air stream is known as the Bernoulli effect.

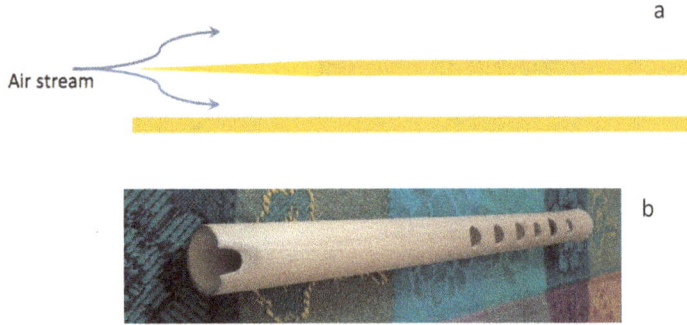

Figure 11.2. (a) Schematic illustration of the air-reed. (b) The sharp edge of a quena flute acts as an air-reed..

vibrations discussed above. The mouthpiece is closed most of the time, opening slightly and briefly as a puff is blown into the bore. Therefore, brass instruments are more like semi-closed pipes, open at one end (the bell) and closed at the other end (the mouthpiece).

A third way to achieve the 'valve' effect is by blowing against a rigid sharp edge, as shown schematically in figure 11.2(a). Upon meeting the sharp edge, the airflow from the mouth can go above or below the non-vibrating edge in an irregular manner. When the airstream flows below the edge, it enters the pipe, and as in the other two excitation mechanisms discussed above, some frequencies will build up a standing wave after a few round-trips in the pipe. As before, the feedback will force the incoming air stream to oscillate above and below the sharp edge at a regular frequency imposed by the standing wave that builds up in the air column. This type of excitation uses the so-called **air-reed**, although there is no reed at all in this case. This excitation mechanism applies to flutes and recorders. Figure 11.2(b) shows the sharp edge on a *quena*, a traditional flute of the Andes. In this case both ends of the pipe are open.

Figure 11.3 shows the playing ranges of some wind instruments. The piano keyboard is shown for reference. We note that each of the instruments shown covers three-plus octaves in different frequency ranges. As discussed above, the flute and piccolo are open on both ends. The other wind instruments shown in the figure are semi-closed. Because the piccolo is shorter than the flute, it has the highest pitch. This can be understood in terms of our analysis of open pipes (see section 6.6) where we found that the mode frequencies in an open pipe are inversely related to the length of the pipe, as expressed by the formula

$$\text{frequency} = (n \times 343)/(2 \times \text{length of pipe})$$

where $n = 1, 2, 3$, etc.

We can use the above formula to estimate the lowest fundamental frequency of a flute, which is 262 Hz (C4). The length of a flute is about 0.66 m. Using the formula, for the fundamental ($n = 1$) we find the frequency 260 Hz, which is in good agreement with the actual frequency of 262 Hz.

From the tonal ranges shown in figure 11.3 we note that the clarinet has lower fundamental pitch than the flute, although the flute is slightly *longer*. The

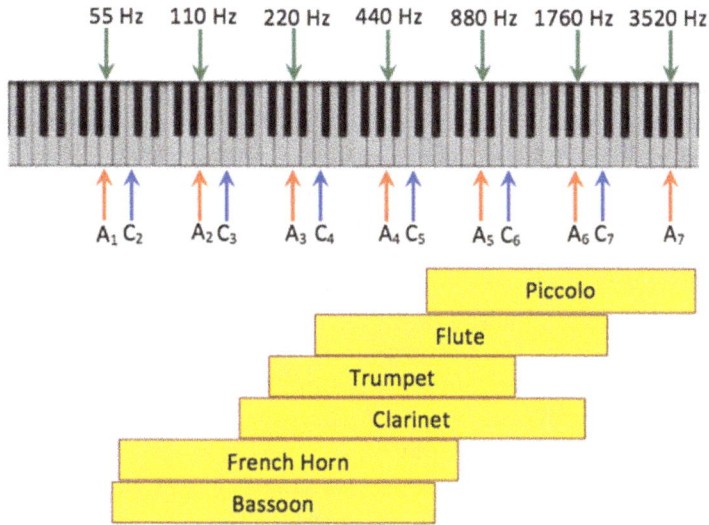

Figure 11.3. Playing ranges of some wind instruments..

Length of Air Column

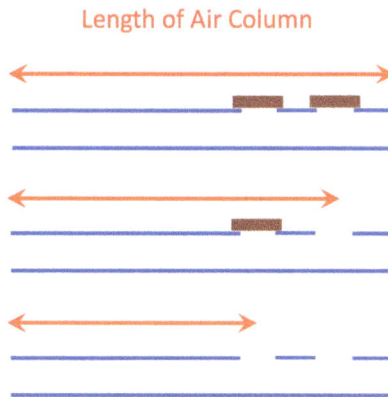

Figure 11.4. Effective length of the air column gets shorter as the holes on the side of the pipe open.

explanation is that the flute is an open pipe while the clarinet is a semi-closed pipe. As discussed in section 6.8, semi-closed pipes can be shorter than open instruments, and produce the same frequency.

The next, and probably most important, question is how do we get the range of notes shown in figure 11.3. According to our model, we need to change the length of the pipe. One way to do this is to open holes at the side of the pipe. Having an open hole on the side of an open pipe in essence changes the 'effective' length of the vibrating air column, as shown in figure 11.4. When the holes are plugged (top) the length of the oscillating air column is essentially the length of the pipe. When the first hole on the right is open, the opening of the hole becomes an open end for the pipe, and the sound escapes from this hole. In other words, the length of the oscillating column or **acoustic length** becomes shorter, and the frequency increases. When the second hole is also open, the acoustic length becomes even shorter, and the frequency higher. Using properly spaced and properly shaped holes at the side of the pipe allows instruments such as the flute and the clarinet to play about three octaves. This tonal range covers about three-dozen tones, which is much more than one can count on their fingers. One way to cover this wide range is to take advantage of harmonics. By blowing hard, or **overblowing,** one can excite the second harmonic, which has twice the frequency of the fundamental, i.e. one octave higher. This works well for the flute, which behaves like an open pipe, where by overblowing one can go from C4 to C5 (refer to figure 11.3 for notes and frequencies), i.e. one octave higher.

For the clarinet, the situation is not as straight forward, because the clarinet behaves like a semi-closed pipe, discussed in section 6.7. In semi-closed pipes the harmonic above the fundamental has three times the frequency of the fundamental. In other words, by overblowing a tone in the clarinet we get more than one octave higher. So overblowing E3 for instance will get us to B4, which is one octave followed by a fifth[3]. Besides overblowing, mechanical keys, fingering, and blowing techniques are used to extend the range to over three octaves and produce all the tones in that range. The specifics are beyond the scope of this book, and the interested reader should refer to more specialized literature.

Much of the above discussion applies to brass instruments, with three main differences. First, instead of using holes to change the acoustic length, brass instruments use valves (e.g. in a trumpet) or sliding pipes (e.g. in a trombone). The principle is schematically illustrated in figure 11.5. Figure 11.5(a) shows an over-simplified setup that can change the effective length of the air-column. When the valve is up (top of figure 11.5(a)) the oscillating air column builds up along the straight section of the pipe. If the valve is down (bottom of figure 11.5(a)) the air column follows the alternative curved path, which is longer. Figure 11.5(b) shows the method used in the trombone. The U-shaped pipe is inside the two straight sections, and can slide back and forth, changing the acoustic length accordingly.

The second difference is that in brass instruments the open end flares into a large bell, which is a significant difference from the straight pipe. As mentioned earlier,

a   b

Figure 11.5. Schematic illustrating the principle of extending the effective length of an air-column by (a) using a valve, and (b) using a sliding U-section.

---

[3] See table 5.2 for definition of intervals.

brass instruments are semi-closed, and one would expect odd harmonics to form (like the clarinet). But because of the bell ending, the mode frequencies change, and in terms of overtones the instrument behaves more like an open pipe; in other words, it can have both odd and even harmonics.

The third difference is that the bell allows a smoother transition of the sound energy from the air column to the surrounding air. The result is that more energy can flow from the instrument, making it sound louder than the flute, etc. In addition, the bell is a wider opening, therefore the sideways spread of the sound is smaller and allows more directionality, especially for higher frequencies (see section 2.6). In other words, the volume and timbre received is very dependent on the direction the instrument is pointing.

11.4 String instruments

String instruments use stretched strings as vibrators. The strings are made of different materials. Earlier instruments used catgut strings (made of sheep intestines). Modern instruments use metals, synthetic materials, or gut. As discussed in section 6.2, the mode frequencies of the standing waves supported by a string depend on the length of the string, and the speed of sound in the string. In turn, the speed of sound in the string depends on the linear density of the string, i.e. the mass per unit length, and on the tension on the string (see section 6.15 for details). If we have two strings of the same material, the thicker string has higher linear density (so-called 'heavy' strings) and the thinner one has lower linear density (so-called 'light' strings). Other things being equal, the heavy string produces a lower frequency than a light string. Also, other things being equal, the higher the tension on the string, the higher the frequency of the tone. The tension that must be applied on the string to produce the right tone depends on the size of the instrument and the type of the string.

Depending on the excitation mechanism, string instruments are further classified into plucked and bowed instruments, although this distinction is not rigid in as much as bowed instruments can be both bowed and plucked. Bowed instruments include the violin, viola, cello, double bass and others. Plucked instruments include the ukulele, mandolin, banjo, lute, guitars and other. The guitars are further classified as classical, steel string acoustic and jazz guitars. We will first describe the parts and function of two of the most common string instruments, namely the violin and the guitar.

Figure 11.6(a) shows the basic parts of a guitar and figure 11.6(b) shows the corresponding parts of a violin. We note several similarities and differences. In both instruments the strings are stretched between the **nut** and the **bridge**. In the guitar, the strings are resting on the saddle (white line in figure 11.1(a)) affixed on top of the bridge. The distance between the nut and bridge defines the **scale length** of the instrument. For a violin the scale length is most commonly about 0.33 m (13 in). For guitars the scale length is most commonly in the range of 0.61–0.66 m (24–26 in). The tension of the strings is adjusted using **tuning pegs**. The violin has four **fine-tuning** pegs on the **tailpiece**. The tuning pegs of the guitar are connected to a gear system (the tuning 'machine') that allows for fine adjustment of the string tension.

The violin has two *f*-**holes** (shaped like the letter *f*) while the guitar has a circular **sound-hole**. Most notably though, the top of the arm or **neck** of the guitar has

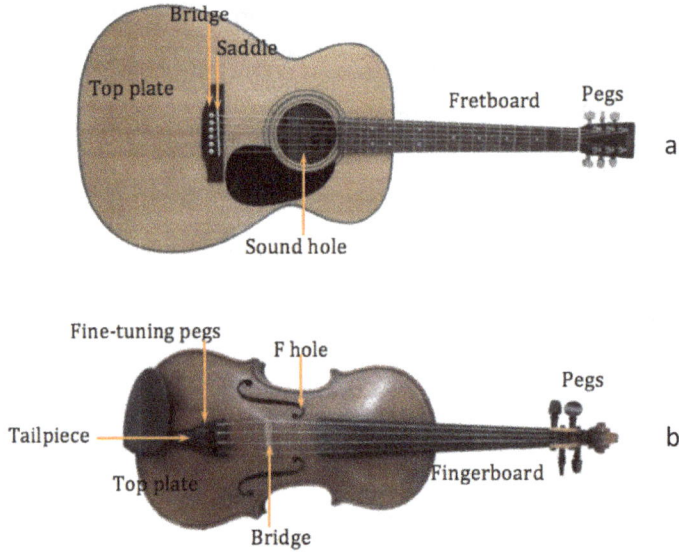

Figure 11.6. Basic parts of a (a) guitar, and (b) violin.

Table 11.1. Standard tuning of open strings for violins and guitars.

Violin					
String	E5 (1st)	A4 (2nd)	D4 (3rd)	G3 (4th)	
Frequency (Hz)	659.3	440	293.7	196	

Guitar					
String	E4 (1st)	B3 (2nd)	G3 (3rd)	D3 (4th)	A2 (5th)	E2 (6th)
Frequency (Hz)	329.6	246.9	196	146.8	110	82.4

regularly spaced metallic frets, forming the **fretboard**, which are absent in the violin's **fingerboard**. The strings are named by the tone they produce when the entire length of the string between the nut and the bridge is vibrating. The string is then said to be 'open'. The strings are also named by numbers (1st, 2nd, etc) starting from the bottom string to top string. The standard tuning of open strings for the violin and guitar together with the note frequencies are listed in table 11.1.

For the violin we see that the frequency ratios of the first two strings is 659.3/440 = 1.498; the same is true for the ratio of the 2nd to 3rd, i.e., 440/293.7 = 1.498; and 3rd to the 4th 293.7/196 = 1.498. The frequency ratio between two notes defines the interval between the two notes. Referring to the discussion of intervals in table 5.2, we see that the interval of 1.498 is the so-called 'fifth'. Hence the violin is 'tuned in fifths'. For the guitar the intervals between open strings are different. For the 1st to 2nd we have 329.6/246.9 = 1.335, which according to table 5.2 is the so-called 'fourth'. The same is true for all ratios except the 2nd to 3rd, i.e. 246.9/196 = 1.260, which is a major third. Hence the guitar is tuned in fourths except for the 2nd to 3rd, which is a major third.

Once the strings are tuned, we can change the vibrating length of the string by 'stopping' the string, i.e. using a finger to press the string against the fingerboard of a violin, or against the fretboard on a guitar. In particular, stopping the string midway between the nut and the bridge reduces the vibrating length to one-half of the open length and as a result, the frequency of this note is double the frequency of the open string, i.e. we get a note that is one octave higher than the open string note. For example, if we stop the 1st string of a violin at the midpoint, we get an E6, i.e. one octave higher than the E5. If we stop the same string somewhere closer to the nut, we will get a tone lower than E6. In principle we can get *any* frequency between E5 and E6 by stopping the E5 string at the right place. Similarly, by stopping the E5 string closer to the bridge, we can get tones higher than E6.

Here of course there are limitations imposed by the length of the fingerboard, and the musician's ability. Usually, the highest playable note on a violin is E7 (or 2637 Hz). The fact that the violin can play any frequency in its tonal range means that the instrument can also play the just scale, and quarter-tones.

The situation is different for the guitar. As in the violin, the midpoint of the open length of the string is one octave higher than the open string. On the guitar, the midpoint (usually marked by two inlays on the fretboard) is at the 12th fret from the nut. In other words, 12 steps on the fretboard correspond to one octave, meaning that the frets correspond to intervals of one semitone.

As discussed in section 5.5 the interval of one semitone in the equal temperament tuning is 1.0595, meaning that moving in steps of one fret from the nut towards the bridge increases the frequency by a factor of 1.0595 or about 6%. Because the frequency is inversely related to the vibrating length (shorter length gives higher frequency) we conclude that moving in steps of one fret towards the nut, the vibrating length of the string increases by a factor of 1.0595, or about 6%. Therefore, the frets are not equally spaced. As the frets are spaced based on the equal temperament scale, the guitar can not play the just scale or quarter-tones.

As strings do not transfer their vibrational energy to the surrounding air efficiently this energy must be first transferred to a box. The box is usually made of fine and carefully selected wood. The thickness of the top plate and back plate is about 2–3 mm (0.08–0.1 in) and both plates can build up standing waves (see section 11.5). But if a system can sustain standing waves of certain mode frequencies, the system can also resonate at the same frequencies. Therefore, the standing wave frequencies that build up on the string can be transferred to the front and back plates of the box, and if the string frequencies match the mode frequencies of the box, we will have resonance. Note that the box does not add energy at all. The energy comes from the vibrating string, and is channeled to the box via the bridge.

In the case of the violin, a sound post inside the box connects the top plate to the back plate and transfers the vibrations to the back plate. The sound post is located under the bridge, on the side of the E5 string (1st string). The sound post also helps support the top plate which experiences a force of about 20 pounds (equivalent to the weight of a mass of 9 kg) that comes from tension force of the strings on the bridge. As an added support, the inner side of the top plate of the violin has a strip of wood that runs under the G3 string (4th string). This so-called **bass bar** not only provides strength to the top plate, but also plays a crucial role in the sound quality of the instrument by controlling the vibration characteristics of the top plate. The guitar body has several support strips on the inner side of the top plate, as well as the back plate for similar reasons. For both violin and guitar, the radiation of low frequencies is primarily from the top plate and the holes. The back plate contributes more to the radiation of the high frequencies.

The violin bow is made of about 150 horsehairs stretched on a stiff stick. Quality bows are made of *pernambuco* wood, although other types of wood and synthetic materials are common. The so-called 'frog' adjusts the tension of the bow. Excitation of the string by the bow occurs because of the friction that develops between the two. In addition, the horsehairs are rubbed with rosin (a material derived from pine trees) to make the bow sticky, and at least momentarily 'catch' the oscillating string, and transfer some energy to the string. This energy exchange depends on the speed of the bow and the force of the bow on the string. If the bow is moving too fast or too slow, or the force of the bow on the string is too high or too low, the resulting sound could be rather unpleasant. The force from the bow that causes the string to vibrate is not continuous. Instead, it follows a 'stick-and-slip' pattern, meaning that the bow momentarily catches the string, then the string slips by the bow, then it catches again, and so on. The rate at which the stick-and-slip occurs is determined by the frequency of the fundamental. In other words, a feedback process is established, as discussed earlier in connection with wind instruments.

As mentioned in section 6.4 the timbre of the tone depends on the point where the string is excited. Thus, bowing the string near the bridge emphasizes the higher harmonics, and produces a louder and more brilliant sound. Bowing closer to the bridge requires higher and steadier force from the bow on the string. Bowing closer to the fingerboard produces a softer sound with less force. Exciting a guitar string is less complicated. As in the case of the violin, the timbre depends on how close to the bridge the string is plucked, but also on the direction of plucking. For example, if the string is plucked parallel to the top plate, the sound is softer and long-lived. If the string is plucked vertical to the top plate, the sound is louder and short-lived.

Figure 11.7 shows the playing ranges of some string instruments. The piano keyboard is shown for reference. Note that the violin has the widest tonal range, almost four octaves. This is remarkable, considering the fact that the violin has two strings less than the guitar.

11.5 Percussion instruments

It is believed that percussion instruments are the oldest, which is reasonable in view of the simplicity that characterizes many percussion instruments. Percussion instruments comprise a wide range of rather dissimilar instruments, their unifying feature being that the sound is produced by striking the vibrator by hand or mallets, or other means. Even within a subcategory, drums for instance, the variations are too numerous to mention. Some percussion instruments use rods or tubes as their

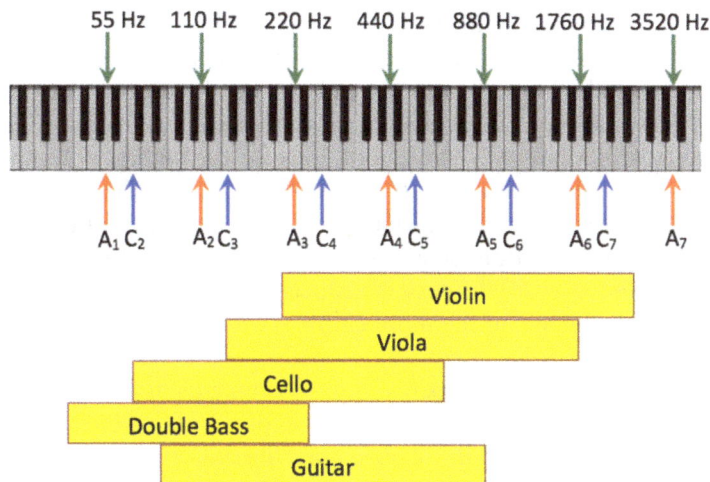

Figure 11.7. Playing ranges of some string instruments.

vibrator, others use stretched membranes, metal plates, etc. The vibrational modes of rods and tubes were described in section 6.10.

Recall that the fundamental produced by a vibrating rod, say an aluminum rod, depends on the length and the thickness of the rod. The frequency of the fundamental mode is lower for a longer rod, i.e. we have the same trend followed by strings. In the case of rods however, the relation between the fundamental frequency and the length of the rod is slightly different. To make the tone one octave higher we need to make the rod four times shorter. Another difference is the relation between frequency and thickness. A thin rod bends more easily than a thick rod, therefore the thin rod will have lower fundamental frequency than a thick rod of the same material and length. This behavior is opposite to what we found for strings. One common characteristic of rods and tubes is that the overtones are not harmonic, i.e. they are not integer multiples of the fundamental.

Chimes or **tubular bells** are essentially a set of 10–20 brass tubes of various lengths, hanging vertically suspended from the top. The top has a protruding rim, which is the striking point. A plastic hammer is commonly used for striking the tubes. The sound we hear is actually the combination of the fourth, fifth and sixth modes, which happen to have frequencies in the ratio 2:3:4. As a result, we hear a missing partial or virtual pitch[4] which is one octave lower than the fourth mode. As the tubes are suspended essentially freely, they can oscillate for a long time, therefore a foot pedal is used to damp out the oscillation.

The **marimba** is a popular instrument that originated in southern Africa. The vibrators of a marimba are horizontal wooden bars suspended near the two ends by strings. The bars are struck with mallets to produce a fundamental frequency, and anharmonic overtones. Under each bar there is a vertical semi-closed pipe. The length of the pipe is selected so that the pipe resonates at the fundamental frequency

---

[4] See discussion of virtual pitch in section 4.10.

of the bar. The rods of the modern marimba are arranged following the pattern of the piano keyboard and the tonal range of a marimba is usually over three octaves.

Recall that for rods the frequency dependence on the length is stronger than in strings. So, if we want to go one octave lower, the length must increase by a factor of four. If we start with a bar of 0.2 m (about 8 in) for the highest note, then the note one octave lower would have a length of 0.8 m (32 in) and the next octave would require a bar 3.2 m (over 10 ft) long, which would be unpractical. For this reason, an arch is carved out of the bottom side of the bars. In this way the bars are thinner at the middle, and the overall stiffness is reduced, leading to lower frequencies of oscillation. The **xylophone** and the more complex **vibraphone** operate essentially on the same principle.

Another important group of percussion instruments uses a two-dimensional surface as the vibrator. The surface can be a sheet of metal or a stretched membrane. As the surface is two-dimensional, we have node lines as opposed to nodes at points. Some of the general trends that are observed in the mode frequencies of strings and rods apply to surfaces as well. For example, the larger the surface, the lower the frequency; the higher the tension on a membrane, the higher the frequency. The mathematical description of two-dimensional modes is complicated, but the pattern of the node lines is rather intuitive. If we have a circular membrane held firmly at the rim, then the rim is a line of nodes, because no vibrational motion takes place at the rim. Figure 11.8 shows the first few modes for a flexible thin membrane. The colors indicate the direction of motion. For example, we can use blue to indicate the upward displacement of the membrane, and red to indicate the downward displacement.

The nodal lines can be straight lines through the center of the membrane or circles centered at the center of the membrane. A pair of numbers labels the modes. The first indicates the number of straight nodal lines; the second number indicates the number of nodal circles. For example, the (0,1) mode has zero nodal lines through the center, and one nodal circle (at the rim). If the membrane moves downward (red) for the first half of a vibration cycle, then it moves upward (blue) in the second half of that cycle, and so on. In the (1,1) mode, we have one nodal line through the center and one nodal circle (the rim). If during the first half of a cycle, the left half of the membrane moves down and the right half moves up, then in the second half of the cycle the motion is reversed. In the (0,2) mode we have no lines through the center and two nodal circles. If the inner part of the membrane is moving up, the outer part is moving down, and the motion pattern alternates every half cycle of the vibration. In the (3,1) mode we have three nodal lines through the center and one circle (the rim). Note that the three lines divide the membrane into six sectors, and that the direction of motion alternates between adjacent sectors. Looking at the other mode patterns, it is clear that the displacement of the membrane changes direction (up–down) as we cross any of the nodal lines (straight lines, or circles).

The frequencies corresponding to the modes shown in figure 11.8 are listed in table 11.2. Here we use the frequency of the (0,1) mode as our unit. The exact frequency of this mode will of course depend on the size and tension of the membrane. But suppose for example that it is 200 Hz. Then the frequency of the (1,1) mode is $1.59 \times 200 = 318$ Hz. In other words, table 11.2 lists frequency ratios.

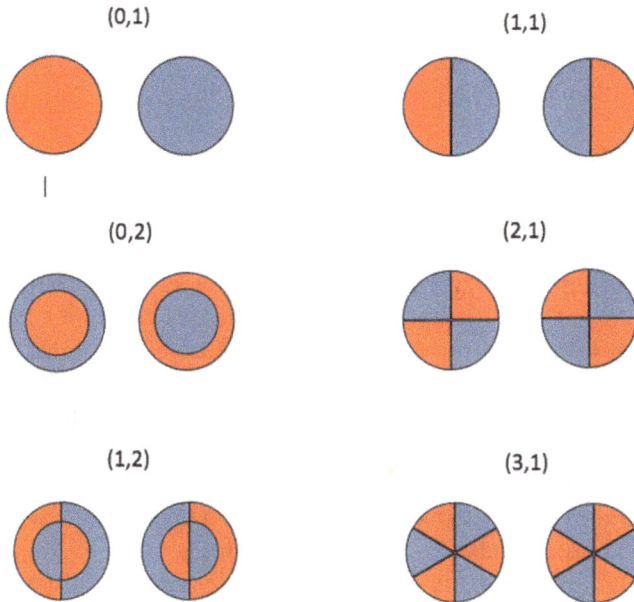

Figure 11.8. First few modes for a flexible thin membrane.

Table 11.2. Frequency ratios corresponding to the vibrational modes shown in figure 11.8.

Mode	(0,1)	(1,1)	(2,1)	(0,2)	(3,1)	(1,2)
Frequency (Hz)	1.00	1.59	2.14	2.30	2.65	2.92

We see that the frequencies of the modes are not integer multiples of a fundamental, i.e. the overtones are anharmonic.

As we move to higher mode numbers, we have more nodal lines; in other words, the surface of the membrane is divided into more sections. Being smaller, these sections can move faster, hence higher mode numbers will generally correspond to higher frequencies, although the exact dependence of the frequency on the mode numbers is not straightforward. The frequency ratios listed in table 11.2 apply for a totally flexible thin membrane with no significant interaction with the surrounding air. Actual drumheads have some stiffness, and some thickness is necessary in order to provide strength. As a result, in an actual drumhead the frequencies of the higher overtones are larger than the values in table 11.2. Also, the interaction with the surrounding air slows down the vibration especially for the lower modes; hence the frequencies of the lower modes are lower than the values in table 11.2. As a result, the frequency of the (1,1), (2,1), (3.1), and (4,1) modes come very close to being a harmonic series, and the sound of the drum has a discernible pitch.

As is the case of string instruments, the quality of the tone depends on how and where the membrane is struck. Looking at the mode patterns, we see that all the modes having nodal lines through the center have a node at the center; therefore, if the membrane is struck at the center, the modes having circular nodal lines will be emphasized. On the other hand, striking near the rim emphasizes the modes having nodal lines through the center.

As the membrane is moving up—for example, in the (0,1) mode—it creates a pressure high. At the same time, the backside of the membrane is creating a pressure low. Therefore, the sound waves created from the front and the back of the membrane are completely out of step. Because of diffraction, the sound bends, so eventually the sound from the back bends around and meets the sound from the front. As the two waves are 180 degrees out of phase, significant cancellation takes place (see section 1.3). The result is a weak sound. To avoid cancellation, drumheads are mounted on baffles that prevent some of the sound from the backside reaching the front, or at least make that sound wave take a longer path, so that the phase difference between the two waves is less than 180° and the cancellation becomes less significant. The membranes in drumheads need to be stretched uniformly, and the tension must be such as to produce the right pitch. This is achieved by using five or more adjustable tension rods to achieve proper tuning.

A typical drum set, shown in figure 11.9, usually includes a **bass drum**, a **snare drum**, and three **tom-toms.** The bass and the snare drum have two drumheads one on each end of their cylindrical body, the shell. The tom-toms can have one or two heads. A set of gut or metal wires is stretched across the lower drumhead, giving the snare drum a rattling sound. Note also that the drum set includes a number of brass conical plates, the **cymbals**, with a dome at the center. The vibrations of a cymbal plate are similar to those of a membrane. In the case of plates, the rim is usually free, therefore it is not a nodal circle. The basic cymbals are the **hi-hat,** the **ride** cymbal, and the **crash** cymbal. The hi-hat consists of two plates mounted on a stand, and is operated by pedal. The ride cymbal is a single plate mounted on a stand. It is struck by drumstick, and has more sustained sound. The crash cymbal is smaller in size than the ride cymbal. It is struck, usually once and hard, as opposed to the ride cymbal that is struck rhythmically.

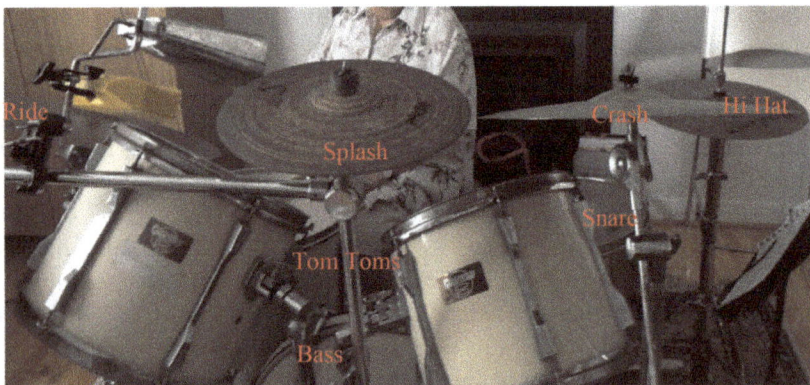

Figure 11.9. A typical drum set.

11.6 Synthesizers

In chapter 3 we discussed how to get the frequency spectrum of a sound. We also compared the frequency spectra of several instruments, and the rate at which the sound builds up and dies out. Having all this information about an instrument, say a classical guitar, one may reverse the process and create the sound of a classical guitar by putting together, or **synthesizing**, the sound from its frequency components. Attempts to synthesize musical sound date back to the late 1800s. In a way, the progress that led to modern electronic synthesizers is parallel to the development of sound recording and reproduction discussed in chapter 10. In both cases, the early stages involved large and expensive devices, and in the modern version, the devices are compact and very affordable.

As with musical instruments, the essential part for making a sound is a vibrator that can produce a range of fundamental frequencies and overtones. In a synthesizer, the frequencies are produced by electronic **oscillators**, i.e. devices that produce voltages of selectable frequency. One can have a large number of such oscillators, one to produce the fundamental, and many others to produce the overtones. This method is called additive synthesis. One could also have oscillators that produce complex tones of selectable fundamental. The complex tone is selected to have a wide range of overtones. The next step is to use filters (see section 8.3) to pass the frequencies needed for producing the desired tone and block the unneeded frequencies. This method is subtractive synthesis. Either way, additive or subtractive, the result is the amplitude spectrum of a musical sound. The next step is to make this sound build up and die out the same way the musical instrument does. This is what we called the 'envelope' in section 3.7. The envelope consists of attack, decay, sustain, and release (ADSR), as shown in figure 3.14. For example, to produce a guitar-like sound, the attack must be quick and pronounced; to produce a flute-like sound, the attack must lead gradually to the sustain part. This can be achieved by amplifying the musical sound at the proper rate, and is done by the so-called **envelope generator**. To produce tonal effects such as vibrato, a separate low frequency oscillator is used to *modulate* (see sections 1.4 and 1.6) the amplitude and frequency of the musical sound.

Clearly all these adjustments cannot be done manually while trying to play a piece of music at the same time. In modern synthesizers, all the tasks described above, adding or subtracting frequencies, filtering, etc, are voltage controlled. For example, the keyboard of the synthesizer produces voltages. Ascending by one key increases the voltage output of the keyboard. This voltage increase is electronically translated to a frequency increase in the fundamental of the oscillator. The increments in the voltage from the keyboard are selected such as to increase the fundamental of the oscillator by one semitone. Similarly, a switch on the synthesizer can select the envelope. Note that one can use the synthesizer to mimic sounds of known instruments and can also produce new sounds that do not correspond to the sound of any musical instrument. One can produce and store a musical piece using a synthesizer. Stored pieces of music can be exported or imported to and from other digital devices. Devices can exchange commands or execute commands that allow one digital unit to control what information is exchanged and when to do so. To enable this exchange, it is necessary to have a communication protocol, i.e. a common language that all digital instruments and computers can understand. One of the most widely used protocols is the Musical Instrument Digital Interface, or **MIDI** for short. The MIDI can be a separate unit within a keyboard that allows the user to enter commands; for example, to tell a synthesizer what to play and when. It could also be in the form of software installed on a computer.

The increased availability of digital memory allows one to store the sounds of musical instruments, or entire sequences of musical sounds, in digital form. The so-called **sampler** is essentially a library of sounds of different instruments. Using the instrument sounds stored in the sampler eliminates the need to synthesize these sounds. The sampler can be a separate unit that stores the sounds in its memory, or the sound library can be stored in computer memory. Either way, using commonly available software (and open-source software) one can use the computer as a synthesizer. Such software allows the user to record and edit sounds and sound sequences to create a musical piece. In addition, by creating different tracks for different instruments one can effectively create an entire music band.

11.7 Further discussion

Classification of musical instruments

The most widely used classification is the Hornbostel–Sachs system, which classifies musical instruments into five main categories, namely **idiophones, membranophones, chordophones, aerophones**, and **electrophones**. Idiophones include instruments such as chimes, cymbals, marimbas, maracas, and others. What these instruments share in common is that the sound is produced by striking, rubbing, or shaking the instrument. In our discussion we referred to these instruments as percussion instruments, with one major exception: the drums. Drums are classified not as idiophones. Instead, they are classified as membranophones, i.e. a vibrating membrane produces the sound.

Besides the string instruments discussed above, chordophones include a variety of other instruments based on vibrating strings, such as the harp, the harpsichord, and the piano. Some authors refer to the piano as a percussion instrument. In a symphonic orchestra, the piano is usually a solo instrument, not part of the string section. On the other hand, the piano has more strings than any other instrument (over 240 strings in a grand piano) but the strings are struck with mechanical hammers, hence the percussion characterization. Aerophones include all the wind instruments discussed earlier, and many more, such as the accordion, the harmonica, and of course the pipe organ, the king of instruments according to some. The pipe organ is a grand collection (over 10 000 in some cases) of open and semi-closed pipes. Some of the pipes use reeds (similar to the saxophone) and others use air-reeds (similar to the flute).

Finally, the electrophones are essentially instruments that use electricity to produce or amplify the sound. This category includes synthesizers, drum machines, and other instruments that use electronic oscillators to produce tones. On the other hand, it is common to attach a microphone to a violin or a guitar, and amplify the sound. In such case the violin or guitar, and many other instruments would be considered electrophones. In other words, there are some unavoidable ambiguities as is the case with most classification systems. After all, it took a decision from the US Supreme Court to classify the tomato as a vegetable rather than a fruit!

11.8 Questions

1.   (a) Which of the following instruments has a vibrator?
         Guitar   Clarinet   Cymbal   Synthesizer
     (b) For the above listed instruments, name the vibrator, if any.
2. Name the radiator for the instruments listed below:
         Guitar   Clarinet   Cymbal   Synthesizer
3. Which of the following instruments has harmonic overtones? Explain your answer.
         Guitar   Clarinet   Cymbal   Synthesizer
4.   (a) What is the 'scale length' of a string instrument?
     (b) Which of the string instruments represented in figure 11.7 can play the lowest notes?
     (c) Which of the string instruments represented in figure 11.7 has largest scale length?
     (d) Which of the string instruments represented in figure 11.7 has the widest tonal range?

Bibliography

The following five books are among my favorite in terms of pedagogical approach, and require minimal mathematical background:

Backus J 1977 *The Acoustical Foundations of Music* 2nd edn (W W Norton)

Moravcsik M J 1987 *Musical Sound: An Introduction to the Physics of Music* (Paragon)

Rigden J S 1985 *Physics and the Sound of Music* 2nd (New York: Wiley)

Campbell M and Greated C 1994 *The Musician's Guide to Acoustics* (Oxford: Oxford University Press)

Berg R E and Stork D G 2005 *The Physics of Sound* 3rd edn (Pearson)

On sound equipment, a very readable and rather accessible book in terms of mathematical and physics background can be found in:

Johnson K W, Walker W C and Cutnell J D 1994 *The Science of Hi-Fidelity* 3rd edn (Kendall/Hunt)

Readers with a more advanced mathematical background, will find a wealth of information on all the topics discussed in this book, and more in:

Rossing T D, Moore F R and Wheeler P A 2002 *The Science of Sound* 3rd edn (Reading, MA: Addison Wesley)

Finally, for those interested in a more rigorous mathematical description of musical instruments, I would suggest:

Fletcher N H and Rossing T D 2005 *The Physics of Musical Instruments* 2nd edn (Berlin: Springer)

Musical Sound, Instruments, and Equipment

Panos Photinos

Appendix A

Scientific notation and abbreviations

1	10^0
1000	10^3 kilo (k)
1 000 000	10^6 mega (M)
1 000 000 000	10^9 giga (G)
0.001	10^{-3} milli (m)
0.000 001	10^{-6} micro (μ)
0.000 000 001	10^{-9} nano (n)

Units
International units (SI) used:

m	meter
s	second
ms	millisecond (1 thousandth of a second)
Hz	Hertz
kg	kilogram
dB	decibel
W	Watt
N	Newton

Appendix B

B.1 Symbols

The symbols used here in connection with waves and oscillations are listed below, with corresponding SI units in parentheses:

c = speed (m s^{-1})
λ = wavelength (Greek letter lambda, m)
f = frequency (Hz)
A_o = amplitude (Pa for a sound; m for strings)
x = distance coordinate (m)
t = time (s)
F = Tension of spring (N)
d = linear density (kg m^{-1})

doi:10.1088/978-1-6817-4680-7app

B.2 Traveling waves

Equation for wave traveling in the positive x direction:

$$A = A_o \cos(2\pi x/\lambda - 2\pi f t) \text{ or} \tag{1}$$

$$A = A_o \sin(2\pi x/\lambda - 2\pi f t) \text{ and more generally} \tag{2}$$

$$A = A_o \sin(2\pi x/\lambda - 2\pi f t + \phi). \tag{3}$$

Here ϕ is the initial phase. If the wave is traveling in the negative x direction, then the negative sign preceding the quantity $2\pi f$ becomes positive. Here the quantity A can be pressure for sound waves or displacement for water waves or strings. A_o is the amplitude of the wave. The above equations give the value of A at distance x away from the origin, and at time t. Note that choice of the origin, and the instant when we started measuring the time, are arbitrary. Recalling that $\cos(0) = 1$, we note from equation (1) that if $2\pi x/\lambda - 2\pi f t = 0$, then $A = A_o$, i.e. we have a crest. The crest moves with speed $c = x/t$. Since $2\pi f t - 2\pi x/\lambda = 0$, we find $x/t = f\lambda$, i.e.

$$c = f\lambda. \tag{4}$$

This is the *phase velocity* of the wave.

B.3 Beats

We will use the trigonometric identity

$$\sin(a) + \sin(b) = 2 \sin\left(\frac{a+b}{2}\right) \cos\left(\frac{a-b}{2}\right) \tag{5}$$

to add two oscillations of equal amplitude and of frequencies f_1 and f_2

$$V \sin(2\pi f_1 t) + V \sin(2\pi f_2 t) = 2V \cos\left(2\pi \frac{f_1 - f_2}{2} t\right) \sin\left(2\pi \frac{f_1 + f_2}{2} t\right).$$

We see that the second factor on the right hand side is an oscillation with frequency equal to the average of f_1 and f_2. The frequency of the cos term is lower.

As an example, we will use $f_1 = 400$ Hz and $f_2 = 402$ Hz, and $V = 1/2$.

The sum of the two oscillations according to the above equations is

$$\cos(2\pi 1 t)\sin(2\pi 401 t)$$

which can be viewed as an oscillation of frequency 401 Hz, with amplitude that varies with a frequency of 1 Hz. As the intensity is proportional to the square of the amplitude, the intensity will get higher when the amplitude reaches its high positive or its low negative value. Therefore, the intensity will oscillate at a frequency of 2 Hz. This is the beat frequency in the example of section 1.4.

B.4 Standing waves in a string

The equation for standing waves is

$$A = A_o \cos(2\pi f t) \sin(2\pi x/\lambda). \tag{6}$$

Here A is the displacement from the equilibrium position of the string. The string is clamped at both ends, therefore, $A = 0$ at $x = 0$ and at $x = L$, where L is the length of the string. Recall that $\sin(n\pi) = 0$ if n is a positive integer $(1,2,3,...)$. For $A = 0$ at $x = L$ we must have $2\pi L/\lambda = n\pi$ or

$$\lambda = 2L/n, \text{ where } n = 1, 2, 3, \text{ etc.} \tag{7}$$

For $n = 1$ we have $\lambda = 2L$ (the fundamental mode), for $n = 2$ we have $\lambda = L$, for $n = 3$ we have $\lambda = 2L/3$ etc, in agreement with figures 6.1–6.3.

The phase velocity of a wave in a string is given in terms of the tension of the string F and its linear density d (i.e. the mass of the string divided by the length if the string)

$$c = \sqrt{F/d} \tag{8}$$

which must equal the product of the frequency times the wavelength. The latter is determined by equation (6), therefore

$$f\lambda = f 2L/n = \sqrt{F/d} \quad \text{or} \tag{9}$$

$$f = \frac{n}{2L} \sqrt{F/d} \tag{10}$$

It follows from equation (10) that:
- The fundamental mode ($n = 1$) has the lowest frequency.
- Longer strings produce lower frequencies.
- The frequency increases with tension.
- The frequency decreases if the linear density is higher, which means that thicker strings (of the same material) produce lower frequencies.

The phase velocity given by equation (4) does not refer to the speed of the up and down oscillation at each point along the string. The speed of oscillation V at each point is

$$V = -2\pi f A_o \sin(2\pi f t) \sin(2\pi x/\lambda). \tag{11}$$

We see that the oscillation speed is different at different points along the string, and also changes with time. The phase velocity refers to the speed of a single wave propagating in very long string.

Standing waves in strings (and pipes) are formed by reflection of waves at the endpoints.

Consider a traveling wave

$$A = (A_o/2) \sin(2\pi x/\lambda - 2\pi ft) \quad (12)$$

and its reflection (i.e. traveling in the opposite direction, as indicated by the + sign preceding the quantity $2\pi ft$)

$$A = (A_o/2) \sin(2\pi x/\lambda + 2\pi ft)$$

(note that here the waves have amplitude $A_o/2$). Adding the two waves and using equation (5), we recover the standing wave equation (6).

Appendix C
Answers to questions

Chapter 1

1.   (a) 120 beats per minute means 2 beats in 1 second, or 0.5 seconds per beat.
      (b) 2 beats per second, or 2 Hz.
2. Speed = Frequency × Wavelength = 20 × 200 = 4000 meters/second
3.   (a) Wave B has higher intensity, because its amplitude is larger.
      (b) The two waves have the same speed.
4. The beat frequency is 106 − 100= 6 Hz

Chapter 2

1.   (a) 3 meters
      (b) 3 meters
      (c) 1.5 meters
2. The speed of sound is higher in hot air. The round trip (echo time) will be **shorter** in hot air.
3. Sound spreads out of the window (see figure 2.5). The volume will be higher when facing the window, but the sound will spread just the same.
4.   (a) time = distance/speed = 3500/343 = 10 seconds.
      (b) 0.00001 seconds. This time is very small compared to 10 seconds, so we can ignore it, and assume that we see the light instantly.
      (c) distance = time × speed = 5 × 343 = 1700 meters (about 1 mile). Hence the rule of thumb, 1 mile for every 5 seconds!

Chapter 3

3.11 Questions

1. 5000 Hz means that we have 5000 positive and 5000 negative peaks in 1 second. So we need at least 10 000 points.
2.   (a) No. It has more than one frequency.
      (b) Yes. It would look similar to figure 1.7, with a beat frequency of 10 Hz.

(c) Two vertical lines: one at 100 Hz, 5 units high, and one at 110 Hz, 2 units high.

(d) No.

(3)   (a) Neither one is a pure tone, because each has more than one frequency.

(b) From figure 3.11 we see that the waveform of the quena flute remains steady for a longer time, therefore the amplitude spectrum of the flute (shown in figure 3.12) will be more or less unchanged over that time interval (the 'sustain' part in figure 3.14). Also, the 'attack' is more gradual for the flute.

4.   (a) Figures 3.2 and 3.4 represent pure tones.

(b) Figure 3.15 (the human voice) contains the largest number of frequency components.

Chapter 4

4.15 Questions

1.   (a) No. The dB is essentially 10 × the logarithm of the intensity ratio (listed in middle column of table 4.1). So, the intensity of the whisper is 10^3 and $10 \times \log (10^3) = 10 \times 3 = 30$ dB. Similarly, for the lawnmower $10 \times \log (10^9) = 10 \times 9 = 90$ dB. What we can add are the intensity ratios of the whisper and lawnmower: $10^3 + 10^9 = 1\ 000\ 001\ 000$. The corresponding dB value is $10 \times \log (1\ 000\ 001\ 000) = 10 \times 9.000\ 000\ 4 = 90.000\ 04$ dB which is essentially equal the power level of the lawnmower in dB, i.e. 90 dB. In summary: when we have two sources, we add their intensities not their dB. As a general rule, we do not take the sum of dB values. We can take the difference between dB values, as explained in section 4.4.

(b) To get the same power level as a jet engine (intensity ratio one trillion) we need 1000 lawnmowers (intensity ratio one billion each).

(c) Each decade of dB means an increase by a factor of 10. There are 4 decades in 40 dB, therefore $10 \times 10 \times 10 \times 10 = 10\ 000$ or 10^4.

(d) The intensity ratio of 10 lawnmowers is $10 \times$ one billion = 10 billion = 10^{10}. To find the power level in dB, we multiply the exponent (i.e., 10) times 10. The result is 100 dB.

2.   (a) The two sources will emit the same amount of power.

(b) From table 4.2 we see that at 30 Hz the threshold of hearing is 60 dB, therefore at this power level the 30 Hz tone is not audible at all! The threshold for the 1000 Hz is 0 dB, therefore the tone is audible.

(c) No, they are not the same. The power level refers to the rate at which the sound energy is received, whether audible or not, and is a physically measurable quantity. The loudness level refers to our perception of the sound.

(d) The two sources will sound equally loud.

(e) From table 4.4 we see that we need to increase the power level by 10 dB to double the loudness. Note that an increase of 10 dB in power level means a 10-fold increase in the intensity ratio.

3.   (a) No, because the difference in frequency is less than 3 Hz, i.e., less than the just noticeable difference.

     (b) Yes, because of the beats. The intensity will go through cycles every 0.5 seconds (because the beat frequency is 1002–1000 = 2 Hz). Remember that the frequency is the inverse of the period of the cycle.

     (c) In part (a) the tones are not played simultaneously. In part (b) they are played simultaneously, hence the beats.

Chapter 5

5.12 Questions

1.   (a) The heptatonic scale uses 7 notes in one octave, the chromatic scale uses all of the 12 notes in the octave.

     (b) Using the equal temperament scale, we can transpose to any key, without having to retune.

2.   (a) C-major scale.

     (b) The major third is E, the fourth is F, and fifth is G.

     (c) The A-minor scale.

     (d) The minor third is C, the fourth is D, and fifth is E.

3.   (a) The present standard is sharper (higher frequency) than the baroque standard.

     (b) The ratio of the frequencies is 440/415= 1.060

     (c) One semitone interval in the equal temperament scale is 1.0595 (see table 5.1)

     (d) They are about 1 semitone apart.

4.   (a) The A-major scale is: A, B, C♯, D, E, F♯, and G♯

     (b) The natural C-minor scale is: C, D, E♭, F, G, A♭, and B♭

Chapter 6

6.16 Questions

1.   (a) The longest wavelength (the fundamental) in the open pipe is two times the length of the pipe, that is 1.2 meters.

     (b) The longest wavelength in the clamped string is two times the length of the string, that is 1.2 meters.

     (c) No. For a given wavelength, the frequency is determined by the speed of sound in the medium, by the relation frequency = speed/wavelength. In the pipe, the speed of sound is about 343 m s^{-1}. In the string, the speed of sound depends on the tension of the string and the properties of the string. This is why in a guitar, although the strings have equal open length, they produce different tones.

2.   (a) The frequency of the overtones will be integer multiples of 50 Hz; so, their frequencies are: 100, 150, and 200 Hz.

        (b) For the semi-closed pipe the frequencies are odd multiples of the fundamental, so the frequencies are: 150, 250, and 350 Hz.

        (c) The open pipe is longer. The length of the open pipe is one-half the wavelength of the fundamental. The length of the semi-closed pipe is one-fourth the wavelength of the fundamental. The wavelength of the fundamental is the same for both pipes in this case therefore the open pipe is longer.

3.   (a) The fundamental is 100 Hz

        (b) The overtone frequencies are 200, 300, 400 Hz and so on.

        (c) The overtones are harmonic because they are integer multiples of the fundamental.

        (d) The frequency of the second harmonic is 200 Hz.

        (e) The second overtone frequency is 300 Hz.

4.   (a) The overtones are not harmonic because they are not integer multiples of the fundamental.

        (b) This is not a pure tone because it contains more than one frequency.

        (c) All the frequencies, i.e., 100, 202, 303, 404,...Hz, are partials of this tone.

Chapter 7

7.6 Questions

1.   (a) The decimal number 6 in binary is 110.

        (b) The binary number 0101 in decimal is 5.

2. The voltage of the wall outlets is analog, because it takes all values between -170 and + 170 Volts (see footnote in section 8.2 for voltage values)

3. All of the input/output ports shown are for analog signals. Some sound cards have a joystick port, which could also be used for MIDI (see section 11.6). This is a digital port. Most MIDIs now have a USB adaptor. The USB is a digital port.

4. Digital computers (there are analog computers as well, but not so common) are based on transistors working essentially as switches. As a switch the transistor can have only two states: ON or OFF. This matches the number of digits used in the binary system: 0 or 1.

Chapter 8

8.10 Questions

1.   (a) According to Ohms Law, the lower impedance (5 Ohms) draws more current from the source.

        (b) Power is the product of voltage times current. As the voltage is constant, the load that draws more current (the 5 Ohms) draws more power.

(c) A thick wire has very low impedance, and will draw the highest current (and power) the voltage source can afford. In this is the case we are 'shorting' the source, which usually means that either the wire will burn out or the source will blow a fuse, or be damaged!

2. (a) A cardioid microphone would be more suitable for a singer in a band, because it can pick up more sound from the singer, less from the other instruments/singers, but also allow the artist some motion.

(b) An omnidirectional microphone would be more suitable for a theater stage because it can pick up sound from all directions.

(c) A cardioid or a shotgun microphone.

3. (a) There are 2 decades in 20 dB, therefore the output is 10^2 larger than the input, or 100 Watts.

(b) It would take a total of 200 Watts: 100 W for sound output and 100 W lost as heat.

(c) Of the 100 W fed to the loudspeaker, the sound output will be 10 W and 90 W will be lost as heat. So of the total 200 Watts supplied to the amplifier, only 10 W will become sound. Not very efficient!

(d) A 4 Ohm load draws more power than an 8 Ohm load.

4. Refer to section 8.6 and figure 8.5.

(a) (a) When the volume setting is low (which means that the input signal from the pre-amp is low) we are still in the linear part of the response curve, therefore the non-linearity is not noticeable.

(b) When the volume setting is high (meaning that the input signal from the pre-amp is high) we are beyond the linear part, and the effects of non-linearity are noticeable.

(c) When the volume setting is high, the non-linearity will increase the total harmonic distortion.

(d) Harmonic distortion means that multiples of the fundamental frequency will appear. For high frequencies, say 10 000 Hz, the harmonics, are 20 000, 30 000 etc, which are mostly or totally outside the audible range. For lower frequencies, say 100 Hz, quite a few of the harmonics (200, 300, 400, 500, etc) are in the audible range. Therefore, the nonlinearity will affect mainly the bass, and midrange.

5. (a) There are 10 000 cm^2 in 1 m^2.

(b) Intensity is power/area, or 0.002 W/0.0001 m^2 = 2 W m^{-2}.

(c) This intensity is at the threshold of pain. Conclusion: when using earphones, the volume must be kept low.

Chapter 9

9.13 Questions

1. We use 343 for the speed of sound, and the formula time = distance/speed.

(a) It will take 5/343 = 0.015 seconds.

(b) Round trip is 10 + 10 = 20 meters. It takes 20/343 = 0.058 seconds.

(c) There are 2 reflections in each round-trip. At each reflection the sound is reduced by a factor of (½) therefore, in one round-trip the sound is reduced to (½) × (½) = ¼ and ¾ or 75% is lost.

(d) Here the sound intensity is reduced by a factor of 0.9 at each reflection, therefore 0.9 x 0.9 = 0.81 or the intensity is reduced to 81%, and 19% is lost.

(e) Recall that the reverberation time is the time it takes for the sound level to drop by 60 dB (a factor of one million). The material that absorbs 10% will give longer reverberation time, because it allows more round-trips before the level is reduced to one millionth.

2. We use 343 for the speed of sound, and the formula time = distance/speed.

   (a) Round-trip distance is 34 + 34 = 68. So it will take 68/343 = 0.2 seconds.

   (b) Ignoring absorption by air, after one round-trip the level is reduced to 0.9 x 0.9 = 0.81 or 81% and 19% is lost.

   (c) In the small hall of the previous problem, with 10% absorption, we lose 19% every 0.058 seconds (the round-trip time), while in the larger hall we lose 19% every 0.2 seconds (the round trip time). In other words, we are losing energy at a faster rate in the small hall, which makes the reverberation time shorter. Conclusion: in larger halls, there is longer time between reflections, therefore, the reverberation time is longer.

   (d) A small hall is more likely to have 'intimacy' because all the distances involved are smaller therefore there is very small time lag between arrivals of the reflections.

Chapter 10

10.7 Questions

1. Magnetic tapes are re-writable, less susceptible to damage, have in some cases more storage capacity and are more compact.

2. CDs have less noise, they do not deteriorate (no mechanical contact to the surface) and, depending on the format used (e.g., MP3), can have much higher storage capacity.

3. Semiconductor storage devices have no moving parts (less susceptible to failure), are more compact and have much higher storage capacity.

4. (a) 10 000 positive peaks in 1 second.

   (b) 10 000 edges are needed to record the waveform for 1 second.

   (c) Two full turns are needed to fit 10 000 edges.

   (d) Two full turns per second. One way to increase the storage capacity would be to have the edges closer by making the size of the pits smaller, but this would require focusing the laser spot more tightly, which has its own challenges.

Chapter 11

11.8 Questions

1. (a) They all have vibrators.
    (b) Guitar: string; Clarinet: air column in the bore; Cymbal: the disk itself; Synthesizer: electronic oscillator.
2. Guitar: box; Clarinet: the open end and holes; Cymbal: the disk itself; Synthesizer: loudspeakers.
3. The guitar and the flute produce harmonic overtones, i.e., integer multiples of the fundamental. The cymbal, like all two-dimensional surfaces do not produce harmonic overtones. The synthesizer can produce harmonic or unharmonic overtones.
4. (a) The scale length is the distance between the bridge (or saddle in the case of the guitar) and the nut.
    (b) The double bass can play the lowest notes (lower frequency and longer wavelength).
    (c) Recall that the wavelength of the fundamental in a clamped string is twice the length of the string. In normal tuning, this means that a longer wavelength is generally required for lower notes. So the double bass has largest scale length.
    (d) The violin has the widest tonal range.

www.ingramcontent.com/pod-product-compliance
Lightning Source LLC
Chambersburg PA
CBHW081535220326
41598CB00036B/6448